Nuclear Engineering Division

Core Programming Topic at the 2011 AIChE Annual Meeting

Minneapolis, Minnesota, USA
16-21 October 2011

ISBN: 978-1-61839-734-8

Printed from e-media with permission by:

Curran Associates, Inc.
57 Morehouse Lane
Red Hook, NY 12571

proceedings.com

Some format issues inherent in the e-media version may also appear in this print version.

Printed by Curran Associates, Inc. (2012)

AIChE
3 Park Avenue
New York, NY 10016-5991

Phone: (203) 702-7660
Fax: (203) 775-5177

www.aiche.org

Additional copies of this publication are available from:

Curran Associates, Inc.
57 Morehouse Lane
Red Hook, NY 12571 USA
Phone: 845-758-0400
Fax: 845-758-2634
Email: curran@proceedings.com
Web: www.proceedings.com

TABLE OF CONTENTS

New Nuclear Forensics Education Program and Research At the University of Utah

Tatjana Jevremovic, Ph.D.
Nuclear Engineering Program, The University of Utah, Salt Lake City, UT

Abstract

To enhance sustainable viability for nuclear forensics related disciplines, such as analytical, geological and radio chemistry, nuclear physics and engineering, the University of Utah Nuclear Engineering Program in collaboration with the Departments of Chemical Engineering, Chemistry and Civil and Environmental Engineering, is establishing an interdisciplinary academic program to support and advance education in the field of nuclear forensics science, technologies and methodologies. Specifically, a new nuclear forensics graduate program track will be offered at the University of Utah. New courses will be developed as a synergy between various disciplines and related to the existing or planned research pertaining to nuclear forensics. As such, this new graduate program will represent the first academic nuclear forensics education in State of Utah. Development of the education program in nuclear forensics is directly connected to new research trusts pertaining to: computational nuclear forensics, radio-analytical chemistry, illicit trafficking of nuclear materials, nuclear fuel cycle and nuclear research reactors data base development & analysis for incidents involving nuclear materials and nuclear fuel trafficking.

Rationale for, and Description of, New Graduate Program in Nuclear Forensics at the University of Utah Nuclear Engineering Program

With the grant received from DOE/DHS in fall of 2010 the Utah Nuclear Engineering Program is building a new graduate program in nuclear forensics. The program at the MS and PhD level has been established and approved. The course work has been defined and will be offered starting fall 2012. The four core (required) courses are described as follows.

As described in GAO-09-527R April 30, 2009, the technical nuclear forensics represents "the analysis of nuclear or radiological materials that are intercepted or the radioactive debris and prompt output signals (such as gamma rays) produced by a nuclear event that can contribute to the identification of the sources of these materials and the processes used to create them." The analytical techniques would then determine the nature of nuclear events in case of nuclear device detonation or release of radiological materials by analyzing the radioactive debris samples (*postdetonation nuclear forensics*). Nuclear forensic techniques also could potentially be used to determine the origin of nuclear or radiological materials or devices seized prior to their use in a weapon (*predetonation nuclear forensics*). The U.S. government's predetonation nuclear forensics capabilities have been demonstrated in investigations on seized nuclear material from illicit smuggling operations (GAO-09-527R, 2009).

Predetonation and postdetonation nuclear forensics combines a wide variety of academic disciplines, such as radiochemistry, nuclear engineering and physics, isotope geochemistry, materials sciences and analytical chemistry. Radiochemistry represents the basis for many of the techniques needed in analyzing the radioactive debris from nuclear events. There are less than 300 experts with nuclear forensics skills and expertise in the country directly involved in nuclear forensics analysis; none of these experts basically work full time in nuclear forensics according to the NTNFC survey conducted in 2008. In addition, the majority of these scientists is close to retirement age. Due to decline in academic programs pertaining to the core disciplines of nuclear forensics, it is difficult to replace the retired force in time yet preserve the hands-on experience and knowledge gained from the past nuclear test programs.

In the spring of 2010 the Utah Nuclear Engineering Program in collaboration with Chemistry Department started the course, "*Radiochemistry with Laboratory I*". The course attracted 35 students from various engineering and science disciplines. In spring of 2011 we had 46 students. It is worth mentioning that "*Radiochemistry with Laboratory I*" represents one of the four core courses for newly established Minor in Nuclear Engineering at the University of Utah. This course represents the base to expand into a new graduate level course, "*Analytical Nuclear Forensics*" that will be required as a prerequisite for "*Nuclear Forensics Technologies, Techniques and Methodologies*" course. This course will be focused at potential sources of forensics in fuel cycle. The steps from uranium ore mining to yellow cake production, UF_6 isotope separation, fuel pellet fabrication and fuel element fabrication will be discussed to identify the various points where terrorists are interested in nuclear materials along this processing route for both dirty bombs and nuclear weapons of mass destruction. In addition, chemical and nuclear processing methods used to produce radioisotopes will be discussed to identify the various points where terrorists are interested in these

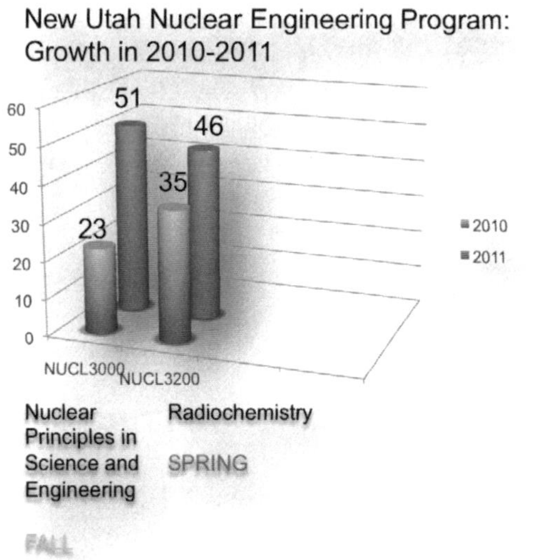

nuclear materials. Special attention will be given to the analytical chemical methods that can be used to measure the composition of nuclear decay products that would be present after a nuclear terrorism incident to determine the ultimate source of the nuclear materials that were used as well as tract the plume of radiation and fall out in the geographical region.

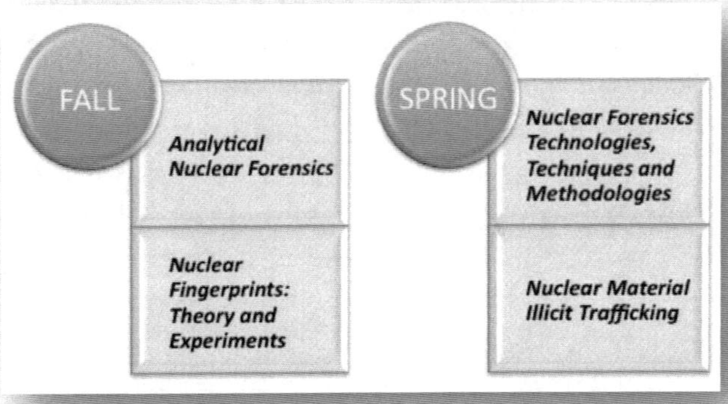

Between 1993 and 2007, there were 1,340 confirmed incidents of illicit trafficking and unauthorized activities involving nuclear and radiological materials worldwide (according to IAEA). Eighteen of these reported incidents involved nuclear material that could be used to produce a nuclear weapon. The worldwide illicit trafficking pathways employ civil infrastructure and particularly the global and domestic transportation networks in different modes such as air cargo, container ships, and freight trucks. The smugglers therefore study such structures in the countries of US and West Europe, for a potentially successful smuggling of nuclear and weapon-grade materials. The transportation sensor network design problem has striking similarities to the predetonation nuclear forensics. The significant similarities include: (1) both systems are complex dynamic spatial systems organized around a hierarchical network structure; (2) both carry flows that vary dynamically, with varying degrees of predictability from origins to destinations; and (3) both have origin-to-destination pairs and the traffic movement can be detected with point or point-to-point detectors. The main differences between the two networks lie in the levels of flow and the consequence of non-detection. Transportation networks see high volumes of flow and errors in demand detection perpetuated by sensors have minimal consequences. Therefore, based on the ongoing research and existing course, both on advancing sensor and traffic network design, we are developing a new graduate course on "*Nuclear Material Illicit Trafficking.*"

The necessity of detecting nuclear materials convertible to weapons (such as U^{233}, U^{235} and Pu^{239}) was recognized as early as the Manhattan Project. When a Congressional Committee asked Dr. J. Robert Oppenheimer how to detect nuclear materials hidden in baggage, he had a simple answer "*use a screwdriver.*" Oppenheimer referred to the fact that well-shielded fissionable materials are almost impossible to detect due to their weak nuclear signatures.

More recently, neutron and gamma interrogation techniques and nuclear fluorescence have provided very exciting opportunities for reliable, non-invasive detection. But, although nondestructive techniques have been developed to detect the fissionable materials, currently installed detection systems do not meet expectations for practical fast and accurate detection. Hence growing efforts are devoted to developing new sensors and sensing systems. Fast and accurate detection of illegally transported nuclear material is crucial in preventing potential terrorist attacks against the US; strengthen the global system of trade; enable new technology and possibly international standards for the global nuclear fuel cycle. Kilograms of fissile plutonium or enriched U^{233} or U^{235} are needed to construct a nuclear bomb. Thus we may assume two general scenarios; one in which small amounts of fissile materials are transferred over a long period of time (in personal baggage) or a larger amount is transported in trucks or containers. Despite growing technological developments, the underlying measurement / detection philosophy remains unchanged representing significant impediment to accurate and fast detection of smuggled nuclear materials. Our current research is focused at the development of fundamentally new approach in detecting weak signals; funded by the NSF/DNDO agency, our computational system development produced a new framework suitable for nuclear detection simulation and fast analysis of complex signals. This project's outcomes are used as a base to develop a new graduate course, *"Nuclear Fingerprints: Theory and Experiments".* The experiments will be developed to emphasize some of the aspects of nuclear signals, i.e. nuclear fingerprints. The course will include computational simulations and the use experiments to benchmark such simulation models.

Establishing New Research in Nuclear Forensics at the University of Utah Nuclear Engineering Program

Although started in late fall of 2010, the team of four professors and a few students by the end of summer of 2011, have been able to produce a number of new results and publish numerous papers (see the references). The following is a brief overview of the new research areas developed and under further development at the Utah Nuclear Engineering Program:

Development of a co-simulation tool for nuclear fuel processing is underway to perform radiochemical plant material and energy balance calculations while at the same time simulating the nuclear criticality. As chemical engineers and nuclear scientists, we recognize that a tool of this type is needed as the nuclear industry moves forward, to avoid accidents, and to save the process engineer time and to save the industry money. With the resurgence of the nuclear industry in the USA there is a need to modernize the tools used for nuclear fuel processing and reprocessing. Today's modern tools for process design include dynamic and steady state mass and energy balance programs like ASPEN Plus and ProSim that make mass and energy balances for a series of unit operations, e.g. reactors, separators of various types that make up a nuclear fuel processing plant. In the design mode this software develops a detailed design of each unit operation giving specific dimensions for the equipment. Some programs, e.g. ASPEN Plus, account for populations of particles as well. The problems with the utilization of these tools for nuclear material processes are that:

1) They lack a physical property data base populated with actinide and lathanide materials properties, however, that can be remedied by purchasing proprietary data bases and tailoring these with experimental data.

2) They are more often than not aqueous based processes while the programs were developed for organic chemical systems. This is not an insurmountable problem but may require proprietary data bases for aqueous species, ions and complexes in solution, like the OLI data base in Aspen for example.

3) They lack physical property data based on isotopes making calculation of unit operations based upon isotope separations impossible.

4) They lack simultaneous calculations for nuclear criticality and radiation dose calculations.

Currently there is not a straight-forward way to analyze the criticality of radiochemical processes in one software package or one that is linked to a mass and energy balance software package. In order to design a chemical and nuclear process accurately, an integrated process simulation tool that has the capability to handle material and energy balances, isotopes, *and* nuclear interactions is needed which is the focus of this is proposed research. The first step is to collect the physical properties of the actinides and lanthanides to populate a proprietary data-base that will allow a process simulator program to perform calculations on actinide processes. The process simulator is ASPEN Plus because it has a dynamic simulator that is needed for ion exchange processes and start up and shut down where most nuclear accidents occur. To perform isotope separations the process simulator will need to be augmented for time dependent physical separation methods. By far the most important improvement in process simulator needed for the nuclear industry is the addition of radiation dose and criticality calculations for all unit operations in a nuclear chemical plant. It must be the focus of future work so that nuclear processes do not have accidents. The parameters needed from the process simulation software to perform criticality calculations are mass, concentration, geometry of the system. The parameters needed for criticality calculations that are not presently in the process simulator property data-base are neutron cross sections as a function of neutron energy and nuclear interactions. The compilation of nuclear cross sections for scattering, absorption, and fission as well as transition probabilities and the energies for the alpha, beta and gamma rays that decay off needs to be created to allow the software to perform criticality calculations and dose prediction. Criticality and dose prediction allows safety analysis to be carried out at the same time as process design to avoid an inadvertent nuclear accident. This can be done by adding this capability to the process simulation software as an additional linked calculation after the mass and energy balance calculations have converged. Criticality calculations involve the solution to the neutron transport equation specifically the effects of fissile material density in the solution on the thermal utilization factor, material and geometric buckling. The nuclear cross sections for scattering, absorption, and fission as well as transition probabilities must be compiled in the physical property data-base as a function of energy. Gamma radiation dose calculations are a direct result of the neutron fluxes that result at the boundary of the equipment. Various numerical approaches to this problem including finite difference, finite element and Monte Carlo calculations have been used in the past. We propose to use Quadrature Method of Moments (QMOM) for this problem as it is an efficient way to solve population balance equations within computation fluid dynamics methods and we have played an important role in their development, validation, verification and uncertainty quantification for particulate processes. This effort will use this understanding of QMOM and tailor it for the solution to neutron populations as predicted by the neutron transport equation. These solutions can also determine the nuclear energy produced that should be coupled back into the process energy balance.

Development of models, algorithms and tools for detecting and interdicting source-to-target patterns of nuclear material illicit trafficking. In order to ship the nuclear materials from a source location with special nuclear material to a target city, the smugglers have to employ the global and domestic transportation network in different modes such as container ships and freight trucks. We will study the radiation sensor network design problem based on an interdisciplinary approach from multiple fields including: nuclear physics, nuclear engineering, transportation engineering, optimization theory, and information theory. An information-theoretic framework will be developed to assist the decision-maker in optimizing the network-wide information gain, system reliability, and total operating cost, by deploying and networking fixed and mobile radiation sensors at different critical locations. The open-source software tools is under development to provide other teachers and researchers with free simulation packages that they can further develop and improve to better fit their needs. In particular, the following critical and theoretically challenging issues will be addressed:

1) How to locate additional mobile or re-locatable radiation detection sensors to improve system redundancy;

2) How to quantify the system-wide economic and risk impacts of miss rates and false alarm rates from different types of existing and new sensors;

3) How to design mobile sensor monitoring routes inside a container terminal to enable collaborative detection and spatial coverage of the sensor nodes.

The integrated simulation framework under development will be able to estimate the error range for each source-to-target flow pair, and their possible error correlation, through a large number of complex but realistic simulated instances under a given sensor network design plan. Different predicted smuggling flow patterns will be considered to ensure the robustness of the suggested sensor location plan. The simulation model is envisioned to be built on a high-performance computational environment to facilitate the most efficient use of emerging parallel hardware.

Based on the known **iMASS** (intelligent Model-Assisted Sensing System) project funded by the NSF/DNDO from 2007 to 2010we are developing an innovative methodology for fast and accurate interrogation of nuclear materials. The novelty of the **iMASS** approach originates from the synergism of smart and customizable Monte-Carlo simulation, advanced signal processing and intelligent information management.[1] The **iMASS** project was executed with great successes. Major goals set in the original scope of work have been accomplished and resulted in some very valuable findings, the most significant one being the ultra upgradability and extensibility of **iMASS** approach. The **iMASS** emerged as an innovative algorithm based on an intelligent information management engine that interactively and adaptively synchronizes the sensing elements with signal processing units to achieve fast and accurate detection of nuclear materials. After three years of research&development, **iMASS** project culminated in reaching its original goal of demonstrating the capabilities using Nuclear Resonance Fluorescence (NRF) as a test bed set of signals. In essence, **iMASS** integrates the following three components into a powerful, accurate yet timely efficient detection system:

- ***Advanced signal processing*- iMASS** consists of a set of advanced signal processing algorithms based on statistical and heuristic methods for detection with quantifiable probability and confidence level[2 3].
- ***Smart Monte-Carlo simulation*- iMASS** integral part is a smart Monte-Carlo simulation package based on GEANT4 capable of reconfiguring itself in response to user's demands[4 5].
- ***Intelligent information management*- iMASS** is supported by a highly sophisticated intelligent information management engine, which collects and integrates measurement data from detectors and performs an anticipatory decision making[6].

The expanded research includes two themes: upgrading **iMASS** for various applications, and extending the scope of **iMASS** by applying it to material tracking. **iMASS** represents a novelty concept that emphasizes and coordinates collaborative efforts among components to maximize performance at system level. Based on this philosophy, **iMASS** is designed as an open architecture in

---

[1] Perry, J., Xiao, S., & Jevremovic, T. (2009). iMASS: Evolved NRF Simulations for More Accurate Detection of Nuclear Threats. In *Proceedings of the 17th International Conference on Nuclear Engineering*. Brussels, Belgium.

[2] P. Forsberg, V. Agarwal, J. Perry, R. Gao, L. H. Tsoukalas and T. Jevremovic. PEAKSEEK: A Statistical Processing

[2] P. Forsberg, V. Agarwal, J. Perry, R. Gao, L. H. Tsoukalas and T. Jevremovic. PEAKSEEK: A Statistical Processing Algorithm for Radiation Spectrum Peak Identification. *21st Int. Conf. Tools of Artificial Intelligence, New Jersey, 2009.*

[3] A. Pantelopoulos, M. Alamaniotis and T. Jevremovic, S. Chung, S. Park, N. Bourbakis. LG-Graph based Detection of NRF Spectrum Signatures: Initial Results and Comparison. *21st Int. Conf. Tools of Artificial Intelligence, New Jersey, 2009.*

[4] Perry, J., Xiao, S., & Jevremovic, T. (2008). iMASS: Computational NRF Spectra Signal from Geant4. In *2008 20th IEEE International Conference on Tools with Artificial Intelligence* (pp. 541-546). Dayton, OH: IEEE. doi: 10.1109/ICTAI.2008.94

[5] Young, J., Perry, J., & Jevremovic, T. (2010). LightGrid - An Agile Distributed Computing Architecture for Geant4. *Nuclear Instruments and Methods in Physics Research Section A: Accelerators, Spectrometers, Detectors and Associated Equipment*, 614(1), 154-158.

[6] Alamaniotis, M., Young, J., Perry, J., Xiao, S., Agarwal, V., Forsberg, P., et al. (2009). Engineering Solution to Nuclear Material Detection at Ports: Introducing the Novel iMASS Paradigm. In *Proceedings of 21st IEEE International Conference on Tools with Artificial Intelligence*. Newark, NJ.

which every component has ample room for further improvements. When a facility owns any type of nuclear material, it is important to be able to track these materials in the interest of safety and security for the facility and the surrounding area. It is crucial to make sure that a facility can track these materials both in transport between facilities (as the NRC tracks) and to track within the facility itself. Part of this would be tracking how moving these materials could affect the people in the area during and after the transport of the material. The effects of the movements/tracking/trafficking of such materials is now analyzed in terms of radiation level change in predefined environment. We consider certain amounts of typical nuclear materials that move (scenario of insiders intending to take these materials out from the predefined environment) and show the spatial change of the radiation levels changing in time in regard to the known background.

References

1. Przybyla, J., Taylor, J., and Zhou, X. (2010) Locating Sensors for Detecting Source -to-Target Patterns of Special Nuclear Material Smuggling: A Spatial Information Theoretic Approach. Sensors. 10(9), 8070- 8091. published, 08/27/2010. http://www.mdpi.com/1424-8220/10/9/8070/
2. Schwerdt, I., Levinthal, J., Dances, C., Choe, D., Yang, H., and Jevremovic, T. (2011) Developing Experimental and Computational Nuclear Forensics Signals of Fast Neutron Irradiation Facility At TRIGA Reactor, in Proceedings of the 2011 AIChE Annual Conference, American Institute of Chemical Engineers, http://www.aiche.org
3. Voyles, A., Yang, H., and Jevremovic, T. (2011) GEANT4 Simulation of Irradiation Facilities and Neutron Sources At University of Utah TRIGA for Nuclear Forensics and Detection, in Proceedings of the 2011 AIChE Annual Conference, American Institute of Chemical Engineers, http://www.aiche.org
4. Taylor, J., Yang, H., Zhou, X., and Jevremovic, T. (2011) GEANT4 Model for Nuclear Forensic Analysis of Nuclear Detonation In Urban Environments, in Proceedings of the 2011 AIChE Annual Conference, American Institute of Chemical Engineers, http://www.aiche.org
5. Sherman, T., Yang, H., Choe, D., and Jevremovic, T. (2011) Monte Carlo Simulation of Tritium Generation Signals In the Utah TRIGA Reactor for Nuclear Forensics, in Proceedings of the 2011 AIChE Annual Conference, American Institute of Chemical Engineers, http://www.aiche.org
6. Andrew Voyles, Haori Yang and Tatjana Jevremovic, GEANT4 – Based Modeling of Research Reactor Irradiation Facilities for Nuclear Forensics, Presented at the 2nd National Conference in Advancing Tools and Solutions

for Nuclear Material Detection, Salt Lake City, Utah, May 2-3, 2011.

7. Todd Sherman and Tatjana Jevremovic, Role of Neutron Activation Analysis in Nuclear Forensics Program at the University of Utah, Presented at the 2nd National Conference in Advancing Tools and Solutions for Nuclear Material Detection, Salt Lake City, Utah, May 2-3, 2011.
8. Greg Moffitt, Charles Grissom and Tatjana Jevremovic, Radiochemistry for Nuclear Forensics at the University of Utah, Presented at the 2nd National Conference in Advancing Tools and Solutions for Nuclear Material Detection, Salt Lake City, Utah, May 2-3, 2011.
9. Jeffrey Taylor, Xuesong Zhou, Haori Yang and Tatjana Jevremovic, Nuclear Forensics in Urban Settings with the use of GEANT4, Presented at the 2nd National Conference in Advancing Tools and Solutions for Nuclear Material Detection, Salt Lake City, Utah, May 2-3, 2011
10. Ian Schwerdt, Joseph Levinthal, Chris Dances, Dong-OK Choe, Haori Yang and Tatjana Jevremovic Experimental Measurements and GEANT4 Modeling of fast Neutron Irradiation Facility at TRIGA for Nuclear Forensics, Presented at the 2nd National Conference in Advancing Tools and Solutions for Nuclear Material Detection, Salt Lake City, Utah, May 2-3, 2011.
11. Taylor, J. and Przybyla, J. (2011) Spatial-Information Approach to Analyzing and Planning Distributed Transportation Security Systems, National Security Innovation Competition Proceedings Report, National Homeland Defense Foundation, published, 05/19/2011. http://www.nhdf.org

(213c) Attenuation of Tetrachloroethylene In Groundwater by Bioremediation

Sathishkumar Santharam[1], Larry Davis[2], Larry Erickson[1] and **Jwan Ibbini**[2], (1)Chemical Engineering, Kansas State University, Manhattan, KS, (2)Biochemistry, Kansas State University, Manhattan, KS

A pilot scale feasibility study was conducted for attenuation of tetrachloroethylene (PCE) contaminated site. PCE concentration in groundwater at the pilot study area was about 15 mg/L (ppm) in the deep zone of the aquifer. Nutrient solutions prepared with different combinations of soy oil methyl esters, cheese whey, yeast extract, lactate and glucose were injected for biostimulation and preconditioning of ground water prior to KB-1 (a consortium of *Dehalococcoides*) injection. Nutrient solution was injected four times during the first year of the pilot study to supplement microbial growth. During biostimulation, no appreciable degradation beyond dichloroethylene (DCE) was evident. The addition of KB-1, during second nutrient injection, reduced PCE and DCE concentrations in the pilot study area to less than 10 percent of the original levels. Total chlorinated ethene concentrations did not reach the original levels two years after the last nutrient addition. The combination of substrates injected provided both short term and long term supply of nutrients for the microbes. Tracer study results demonstrated that microbial growth decreased ground water velocity during the study, but returned to normal conditions one year after the last nutrient addition.

See more of this Session: Contaminant Transport and Site Remediation
See more of this Group/Topical: Environmental Division

Equilibrium and Kinetic Studies on Reactive Extraction of Propionic Acid using Tri-*n*-Octylamine in 1-Decanol + Cyclohexane (1:1 v/v)

Sushil Kumar[1] and B V Babu[2]

[1]Assistant Professor, Chemical Engineering Department, Birla Institute of Technology and Science (BITS), PILANI, Rajasthan, 333 031, India
E-mail: skumar@bits-pilani.ac.in

[2]Director, Institute of Engineering and Technology (IET), JK Lakshmipat University (JKLU), Ajmer Road JAIPUR - 302 026
Phone: +91-141-2168-330 /344 / 167
E-mail: director.iet@jklu.edu.in; profbvbabu@gmail.com

Abstract

The present study is aimed at the recovery of propionic acid from aqueous solution by reactive extraction with tri-octyl-amine (TOA) dissolved in a mixture of decane-1-ol (an active diluent) + cyclohexane (an inactive diluent). The equilibrium and kinetic extraction experiments are carried out to investigate the effect of modifier composition (20 - 80%), the effect of initial acid concentration (0.0675 to 0.675 mol L^{-1}*) and the effect of agitation speed (250 – 500 rpm) on the extraction efficiency at a constant temperature of 298 K. The higher modifier (1-decanol) concentrations lead to an increase in the extraction efficiency (degree of extraction) of acid. The degree of extraction (%) decreases significantly when the concentration of propionic acid is increased with TOA in cyclohexane/1-decanol (1:1 v/v). A mathematical model based on mass action law and a population-based search algorithm (differential evolution, DE) is proposed and used to estimate the extraction equilibrium constants (*K_E*) and stoichiometry of reactive extraction. Individual equilibrium constants for simultaneous formation of (1:1) and (2:1) acid:amine complexes are also determined. The experimental kinetic data for the extraction of propionic acid are interpreted by a forward reaction rate with* $\alpha' = 1.68$ *and* $\beta' = 1.0$ *and the reverse reaction rate with* $\gamma' = 1.0$*, a very good fit is obtained with almost constant values for both the rate constants,* k_1 *and* k_2*, for the whole range of acid (0.27 – 0.54 mol.*L^{-1}*) and a constant concentration of amine (0.457 mol.*L^{-1}*).*

Keywords: *Separation; reactive extraction; equilibria; kinetics; modeling; differential evolution (DE).*

Introduction

Propionic acid is most widely used in the field of food and beverages as an acidulant and also in pharmaceutical and chemical industries [1]. Due to growing demand of pure, naturally produced propionic acid, the interest towards propionic acid recovery from fermentation broths and aqueous effluents is increased. Specific uses of propionic acid draw attention toward a better recovery process, which will increase the productivity and reduce the overall production cost [2]. A reactive extraction process exploits reversible chemical complexation in the extractant phase, provides an effective separation, especially for dilute solutions such as the aqueous solution of propionic acid in a fermentation broth [3-5]. Long-chain aliphatic amines are effective extractants for the separation of carboxylic acids from dilute aqueous solution. The specific chemical interactions between the amines and the acid molecules to form acid-amine complexes in the extractant phase allow more acid to be extracted from the aqueous phase. The specific affinity of long-chain tertiary amines for propionic acid gives high selectivity for this type of solute with respect to water and eventually non-acidic species in the mixture [5-10]. The diluents can also hydrogen bond with the oxygen accessible of the acid:amine complexes. Since the presence of the hydroxyl and carboxylic groups increases the solubility of acids in the water phase, the strong interactions of solvent with solutes are necessary to extract carboxylic acids from aqueous solutions. Polar diluents that enhance the extraction power of amines are more favorable than nonpolar diluents, as shown in the extraction of carboxylic acids [11,12]. The present study is aimed at the recovery of propionic acid from aqueous solution by reactive extraction with tri-octyl-amine (TOA) dissolved in a mixture of decane-1-ol (an active diluent) + cyclohexane (an inactive diluent). The equilibrium and kinetic extraction experiments are carried out to investigate the effect of modifier composition (20 - 80%), the effect of initial acid concentration (0.0675 to 0.675 mol L^{-1}) and the effect of agitation speed (250 – 500 rpm) on the extraction efficiency at a constant temperature of 298 ± 1 K.

Experimental Section

Materials

The materials used in this study are propionic acid, long chain aliphatic amine based extractant [tri-n-octyl amine (TOA)] and diluents [1-decanol (active diluent, a modifier) and cyclohexane (inactive diluent)], as listed in Table 1.

Table 1. Materials used in the present study

S. No.	Reagents	Suppliers	Purity	M. Wt., $g.mol^{-1}$	Spec.gravity $g.cm^{-3}$	Viscosity (cP)
1.	Propionic acid	Sigma-Aldrich, USA	98%	74.08	0.990	-
2.	Tri-n-octyl amine (TOA)	Fluka, USA	98%	353.68	0.809	8.32 (25^{0}C)
3.	1-Decanol	Spectrochem, India	98%	158.28	0.830	34 (22°C)
4.	cyclohexane	Spectrochem, India	99%	84.16	0.779	0.98 (25^{0}C)

Methods

Propionic acid is dissolved in de-ionized water to prepare the aqueous solutions with initial concentration in the range of 0.0675 to 0.675 $mol.L^{-1}$ (0.5 to 5% w/w). The organic solutions are prepared by dissolving extractant (TOA) in the mixture of cyclohexane and 1-decanol. The extraction equilibrium experiments are carried out at constant temperature (298 ± 1 K) with equal volumes (20 cm^3 of each phase) of the aqueous and organic solutions shaken on a temperature controlled reciprocal shaking water bath (HS 250 basic REMI labs) using conical flasks of 100 mL. After attaining equilibrium, the phases are brought into contact with each other for separation. After settling, organic and aqueous phases were separated. The concentration of acid in the aqueous phase is determined by taking a sample of 2 cm^3 volume and using titration with fresh sodium hydroxide solution of 0.02 N and phenolphthalein as an indicator. The acid concentration in the organic phase is calculated by mass balance. The results are found to be reproducible within ±5%.

The effectiveness of the reactive extraction process is analyzed by calculating the distribution coefficient (K_D), the degree of extraction (E) and loading ratio (Z) as given by Eqs. (1), (2) and (3), respectively.

$$K_D = \frac{\overline{C}_{\mathrm{HC}}}{C_{\mathrm{HC}}} \qquad (1)$$

$$E = \frac{\overline{C}_{HC}}{C_o} = \frac{K_D}{1+K_D} \quad (2)$$

$$Z = \frac{\overline{C}_{HC}}{[\overline{T}]_o} \quad (3)$$

Mathematical Modeling

Equilibrium Model

Assuming that *m* molecules of propionic acid (HP) react with *n* molecules of extractant, TOA (N), the extraction equilibrium is represented as:

$$m\text{HP} + n\overline{\text{N}} \Leftrightarrow \overline{(\text{N})_n(\text{HP})_m} \quad (4)$$

The equilibrium constant (K_E) is calculated using Eq. (5)

$$K_E = \frac{[\overline{(\text{N})_n(\text{HP})_m}]}{[\text{HP}]^m[\overline{\text{N}}]^n} = \frac{[\overline{(\text{N})_n(\text{HP})_m}](1+K_a/[\text{H}^+])^m}{[\overline{\text{N}}]^n \text{C}_{\text{HP}}^m} \quad (5)$$

The distribution coefficient is defined as:

$$K_D = \frac{\overline{\text{C}}_{\text{HP}}}{\text{C}_{\text{HP}}} = m\frac{[\overline{(\text{N})_n(\text{HP})_m}]}{\text{C}_{\text{HP}}} \quad (6)$$

Substituting the values of $[\overline{(\text{NR}_3)_n(\text{HC})_m}]$ from Eq. (6) in Eq. (5) results Eq. (7).

$$K_E = \frac{K_D(1+K_a/[\text{H}^+])^m}{m[\overline{\text{N}}]^n \text{C}_{\text{HP}}^{m-1}} \quad (7)$$

The free extractant concentration, $[\overline{\text{N}}]$ in the organic phase, represented as:

$$[\overline{\text{N}}] = [\overline{\text{N}}]_{in} - n[\overline{(\text{N})_n(\text{HP})_m}] \quad (8)$$

$$\Rightarrow [\overline{\text{N}}] = [\overline{\text{N}}]_{in} - K_D n\text{C}_{\text{HP}}/m \quad (9)$$

Using the Eqs. 7 and 9, Eq. 10 results in

$$K_D = mK_E\left([\overline{\text{N}}]_{in} - K_D n\frac{\text{C}_{\text{HP}}}{m}\right)^n \frac{\text{C}_{\text{HP}}{}^{n-1}}{(1+K_a/[\text{H}^+])^m} \quad (10)$$

The values of equilibrium extraction constant (K_E) and the number of reacting acid & extractant molecules (*m* & *n* respectively) can be estimated using Eq. (10).

Based on the estimated values of m per amine molecule, the formation of different types of complexes, i.e., (1,1) and (2,1) are considered. The stoichiometric equations describing the extraction are given by Eqs. (11,12):

$$\mathrm{HP} + \overline{\mathrm{N}} \Leftrightarrow \overline{\mathrm{HPN}} \tag{11}$$

$$\mathrm{HP} + \overline{\mathrm{HPN}} \Leftrightarrow \overline{(\mathrm{HP})_2\,\mathrm{N}} \tag{12}$$

The corresponding extraction constants are calculated using Eqs. (13,14):

$$K_{11} = \frac{[\overline{\mathrm{HPN}}]}{[\mathrm{HP}][\overline{\mathrm{N}}]} = \frac{\overline{\mathrm{C}}_{11}(1 + K_a/[\mathrm{H}^+])}{\mathrm{C}_{\mathrm{HP}}\mathrm{C}_{\mathrm{N}}} \tag{13}$$

$$K_{21} = \frac{[\overline{(\mathrm{HP})_2\,\mathrm{N}}]}{[\mathrm{HP}][\overline{\mathrm{HPN}}]} = \frac{\overline{\mathrm{C}}_{21}(1 + K_a/[\mathrm{H}^+])}{\mathrm{C}_{\mathrm{HP}}\overline{\mathrm{C}}_{11}} \tag{14}$$

$\overline{\mathrm{C}}_{11}$ and $\overline{\mathrm{C}}_{21}$ are the concentrations of the complexes of (1,1) and (2,1) respectively. The acid concentration and free amine concentration in the organic phase are given by Eqs. (15) and (16), respectively.

$$\overline{\mathrm{C}}_{\mathrm{HP}} = \overline{\mathrm{C}}_{11} + 2\overline{\mathrm{C}}_{21} = K_{11}\mathrm{C}_{\mathrm{N}}\,\mathrm{C}_{\mathrm{HP}}\big/(1 + K_a/[\mathrm{H}^+]) + 2K_{21}\overline{\mathrm{C}}_{11}\,\mathrm{C}_{\mathrm{HP}}\big/(1 + K_a/[\mathrm{H}^+]) \tag{15}$$

$$\mathrm{C}_{\mathrm{N}} = \mathrm{C}_{\mathrm{N}_{\mathrm{in}}} - (\overline{\mathrm{C}}_{11} + \overline{\mathrm{C}}_{21}) = \mathrm{C}_{\mathrm{N}_{\mathrm{in}}} - (K_{11}\mathrm{C}_{\mathrm{N}}\,\mathrm{C}_{\mathrm{HP}}\big/(1 + K_a/[\mathrm{H}^+]) + K_{21}\overline{\mathrm{C}}_{11}\,\mathrm{C}_{\mathrm{HP}}\big/(1 + K_a/[\mathrm{H}^+])) \tag{16}$$

The values of the equilibrium constants K_{11} and K_{21} are estimated based on the total acid concentration in the aqueous phase at equilibrium, and by applying the mass action law (experimental results).

Kinetic Model

Chemical reaction between acid molecules and amine molecules is represented by Eq. (4). By analogy with the true elementary reaction, the rate equation may be written as:

$$-r_A = -\frac{dC_{HA}}{dt} = k_1 C_{HA}^{\alpha'} C_S^{\beta'} - k_2 C_C^{\gamma'} \tag{17}$$

where, the individual orders α', β' and γ' are to be determined from the experimental kinetic data. The concentrations of all the components in the rate expression (17) can be derived from Eq. (4) taking the acid concentration in the aqueous phase as a basis for calculations.

Results and Discussion

The results obtained on extraction of propionic acid with TOA dissolved in an inactive diluent and a modifier, are presented. Diluents chosen in the study are cyclohexane from inactive chemical class, and 1-decanol as a modifier from active chemical class to examine the effect of diluent-complex interactions.

Equilibrium Studies

The effect of the concentration of 1-decanol on the degree of extraction of propionic acid (0.405 mol.L^{-1}) is first studied (Figure 1). The organic phase is composed of extractant (TOA), a modifier (1-decanol), and an inert diluent (cyclohexane). The TOA concentration is fixed at 20% v/v (0.46 mol.L^{-1}). The higher concentrations of modifier (1-decanol) lead to an increase in the extraction efficiency (degree of extraction, E) of acid. There is a very strong modifier effect on the degree of extraction, when cyclohexane is used as an inert diluent. No significant change is found in the volumes of organic phase and aqueous phase after their separation. The values of estimated equilibrium constant (K_E) and stoichiometry (n) depends not only on the concentration of modifier in the presence of inert diluent, but also on the volume phase ratio between modifier and diluent [13]. The volume phase ratio between modifier and diluent is taken one in this study. Figure 2 shows the experimental data for the system having cyclohexane as an inert diluent and 1-decanol as a modifier. There is a significant increase in the degree of extraction (E) for an increase in TOA concentration up to 20 %. For the TOA concentration higher than 20 %, an insignificant increase in extraction efficiency is found with cyclohexane/1-decanol mixture of 1:1 v/v.

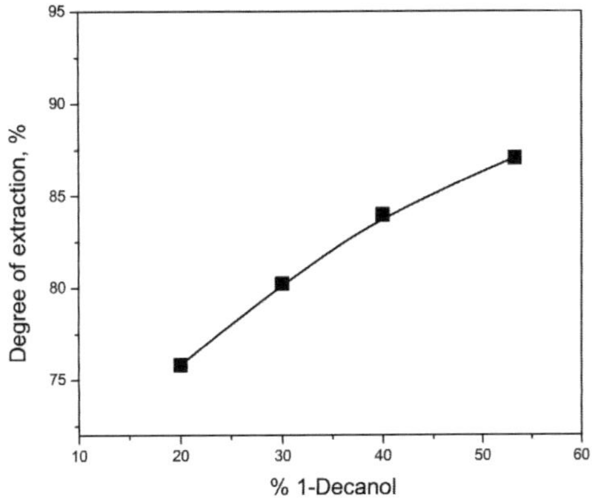

Figure 1 Influence of modifier (1-decanol, %) in the organic phase (20% TOA + cyclohexane + 1-decanol) at initial propionic acid concentration of 0.405 mol.L^{-1}.

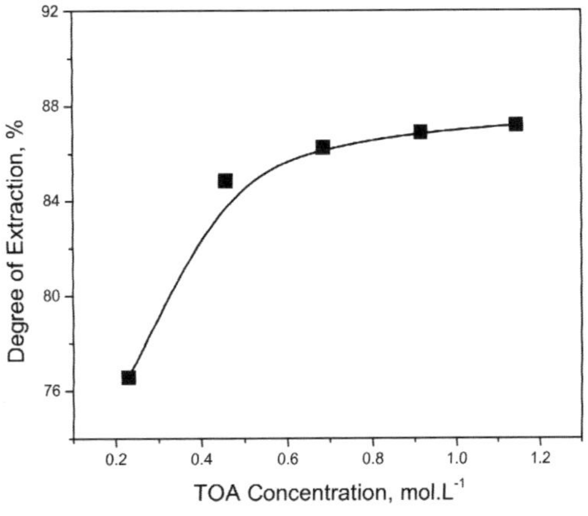

Figure 2 Effect of TOA concentrations in cyclohexane + 1-decanol (1:1 v/v) at constant propionic acid concentration (0.405 mol.L^{-1})

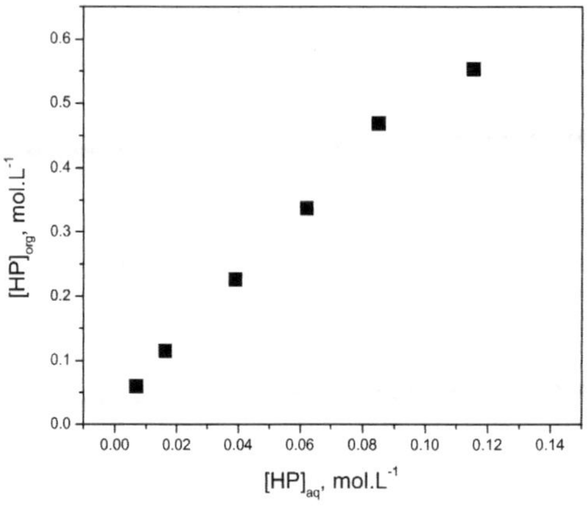

Figure 3. Equilibrium isotherms of propionic acid for TOA concentration (0.46 mol.L^{-1}) in cyclohexane + 1-decanol (1:1 v/v)

The isotherms for propionic acid are obtained from different aqueous solution concentrations (0.0675 – 0.676 mol.L^{-1}), at a constant concentration of TOA (0.46 mol. L^{-1}) dissolved in cyclohexane & 1-decanol (1:1 v/v) and the results are presented in Figure 3. Almost linear relationship is found between acid concentration in the two phases at equilibirum, and slightly nonlinear relationship for higher concentrations due to Henry's law type isotherm is only valid for low acid concentration range. From Figure 3, the degree of extraction (as slope of the data) decreases significantly when the concentration of propionic acid is increased from 0.0675 to 0.675 mol.L^{-1}. At the higher range concentrations of propionic acid, the concentration of extractant (TOA) may be limiting factor for the extraction of propionic acid. The effect of 1-decanol on extraction efficiency is found to be significant with inert diluents. The viscosity of 1-decanol is very high as compared to inactive diluents such as cyclohexane. Hence, the equal volume mixture of cyclohexane and 1-decanol is used to estimate the equilibrium extraction constant (K_E) and the stoichiometries of extraction reaction. 20% of TOA as an extractant composition is used for the extraction of propionic acid in the range of 0.0675 – 0.676 mol.L^{-1}. From Figure 2 (E) and Table 2 (loading ratio, Z), it is observed that E and Z, corresponding to different acid concentrations at constant TOA concentration, reaches towards a plateau. It contemplates a tendency towards the formation of two types of complex structures, i.e., multiple acid per one amine ($n < m$) aggregation, or an equimolar structure ($m = n$). This fact seems to be a common strategy for designing the amine extraction of propionic acid.

Table 2. Equilibrium results for reactive extraction of propionic acid using TOA (0.46 mol.L^{-1}) dissolved in cyclohexane and 1-decanol (1:1 v/v) at 298 K

Diluent system	C_{HP}, mol.L^{-1}	$\overline{C}_{HP}$, mol.L^{-1}	$\overline{C}_{HP}$, pred (Eq.15) mol.L^{-1}	$K_{D,\,exp}$	$K_{D,\,pred}$ (Eq 10)	E	Z	pH_{eq}
Cyclo-hexane + 1-decanol	0.008	0.059	0.048	7.41	5.54	88.1	0.13	4.12
	0.02	0.115	0.117	5.89	6.44	85.48	0.253	3.68
	0.044	0.226	0.248	5.19	6.00	83.85	0.495	3.44
	0.068	0.337	0.356	4.93	5.12	83.13	0.737	3.37
	0.091	0.469	0.451	5.14	4.42	83.71	1.026	3.24
	0.123	0.553	0.516	4.50	3.67	81.83	1.211	3.18

With these experiments, the loading ratios for the complete range of propionic acid concentration, found in the range of 0.113 – 1.05 indicate that 1:1 and 2:1 complexes between acid and TOA are formed simultaneously. The values of equilibrium extraction constant (K_E), the number of acid molecule (m) per reacting extractant molecules are estimated by solving model equation (10) using optimization procedure. A population based search algorithm called differential evolution (DE), which is simple and robust and has a proven successful record for various engineering applications in both single- and multi-objective optimization is employed [14,15]. An objective function based on least square error between experimental and predicted values of K_D is minimized. The estimated values of number of acid molecule (m) per TOA molecule in the formation of acid:TOA (m:n) complexes, and equilibrium extraction constant (K_E) are given in Table 3. The values of K_{11} and K_{21} for simultaneous formation of (1:1and 2:1) complexes, respectively between acid and TOA are also given in Table 3.

Table 3. Values of K_E, the number of reacting acid and extractant molecules (*m and n*), and individual equilibrium constants (K_{11} and K_{21}) for formation of (1:1and 2:1) complexes respectively using DE.

Diluent system	C_{in} (mol.L^{-1})	TOA (mol.L^{-1})	m	n	K_E	K_{11}	K_{21}
cyclohexane + 1-decanol (1:1 v/v)	0.0675 – 0.675	0.457	1.31	1	48.93	13.56	7.56

Kinetic Studies

In order to select the stirring condition, preliminary kinetic measurements are performed with the two-phase system of an aqueous phase. The shape and the position of the evaluated kinetic curves for agitations above 350 min^{-1} is almost the same. Hence, according to these preliminary measurements, the selected stirring speed of 350 rpm excludes the influence of the agitation on kinetics. This means that the effect of mass transfer on the overall kinetics is minimized and presumably the interfacial chemical kinetics can be determined from the experimental results according to Poposka *et. al.,* [16]. Therefore, the kinetic experiments for the reactive extraction of propionic acid using TOA in a mixture of cyclohexane and 1-decanol (1:1 vol%) are carried out with the stirring speed of 350 rpm and systems are allowed to achieve an equilibrium. In this kinetic study, the best pair of rate constants for a given particular set of individual orders - α', β' and γ' in the rate expression (17) are determined. The value of stoichiometric coefficient (m) is found to be 1.31 (Table 3). The different values of individual reaction orders (α', β' and γ') in Eq. (17) are considered to obtain the rate constant k_1 and k_2, but different values for the rate constants are found for the given concentrations (0.27 – 0.54 mol.L^{-1}) of acid at a constant concentration of TOA (0.457 mol.L^{-1}). When the experimental kinetic data are interpreted by a forward reaction rate with $\alpha' = 1.31$ and $\beta' = 1.0$ and the reverse reaction rate with $\gamma' = 1.0$, a very good fit is obtained with almost constant values for both the rate constants (k_1 and k_2). This rate form [$\alpha' = 1.31$, $\beta' = 1.0$, $\gamma' = 1$ in Eq. (17)], describing the net rate of extraction, can be considered as elementary kinetics, which suggests that the chemical complexation of the acid is a multiple reaction mechanism through which several types of acid-amine complexes are formed. The values of the rate constants k_1 and k_2 are presented in Table 4. The evaluated kinetic curves are shown in Figure 4 as solid lines.

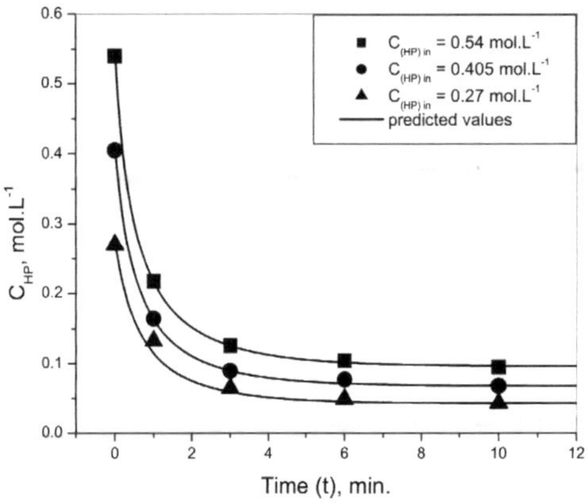

Figure 4 Experimental and estimated kinetic curves for extraction of propionic acid (0.27 - 0.54 mol.L^{-1}) by TOA (0.46 mol.L^{-1}) in cyclohexane + 1-decanol (1:1 v/v)

Table 4. Values of the rate constants k_1 and k_2 in the rate equation (17) for the extraction of propionic acid by TOA (0.46 mol.L^{-1}) in cyclohexane+1-decanol (1:1 v/v)

α' = 1.31, β' = 1.0 and γ' = 1.0			
C_{in}, mol.L^{-1}	k_1	k_2	k_1/k_2
0.54	6.332	0.0858	73.80
0.405	6.874	0.0932	73.75
0.27	6.974	0.0945	73.80
Average value of k_1 = 6.73 and k_2 = 0.0912			

Conclusions

The equilibrium and kinetic studies on reactive extraction of propionic acid by tri-n-octylamine (TOA) dissolved in the equal volume mixture of cyclohexane and 1-decanol is presented for the intensification of recovery of propionic acid from the fermentation broth and waste stream. The extraction efficiency in terms of degree of extraction is found to increase with an increase in the concentration of modifier (1-decanol) and TOA. The loading ratios for the complete range of propionic acid concentration, found in the range of 0.113 – 1.05 indicate that 1:1 and 2:1 complexes between acid and TOA are formed simultaneously. The estimated values of overall K_E, and individual K_{11} for 1:1 complexes and K_{21} for 2:1 complexes between acid and TOA are found to be 48.93, 13.56 and 7.56, respectively. The predicted values of acid concentrations using a model for individual complexes (Eq. 17) using the DE approach correctly represent the experimental values of acid concentrations in the organic phases. In the kinetic study, the extraction of propionic acid is interpreted by a forward reaction rate with α' = 1.68 and β' = 1.0 and the reverse reaction rate with γ' = 1.0 with the corresponding rate constants of k_1 = 6.73 and k_2 = 0.0912.

References

[1] M.J. Playne, Propionic and butyric acids, in: Comprehensive Biotechnology, ed by Moo Young M, Pergamon publishers, New York, 3 (1985).

[2] S. Kumar, B.V. Babu, propionic acid production via fermentation route using renewable sources, Chem. Ind. Digest 9 (2008) 76-81.

[3] J. Hartl, R. Marr, Extraction processes for bioproduct separation, Sep. Sci. Technol. 28 (1993) 805.

[4] K.L. Wasewar, Separation of lactic acid: recent advances, Chem. Biochem. Eng. Q. 19 (2005) 159-172.
[5] S. Kumar, B.V. Babu, Process intensification for separation of carboxylic acids from fermentation broths using reactive extraction, J. Fut. Eng. Technol 3 (2008) 19-26.
[6] C.J. King, L.J. Poole, Regeneration of carboxylic acid-amine extracts by back-extraction with an aqueous solution of volatile amine, Ind. Eng. Chem. Res. 30 (1991) 923-929.
[7] K.L. Wasewar, V.G. Pangarkar, Intensification of propionic acid production by reactive extraction: effect of diluents on equilibrium, Chem. Biochem. Eng. Q. 20 (2006) 1-7.
[8] S. Kumar, D. Datta, B.V. Babu, Experimental data and theoretical (chemodel using differential evolution approach and linear salvation energy relation model) predictions on reactive extraction of mono-carboxylic acids using tri-n-octylamine, J. Chem. Eng. Data. 55 (2010) 4290-4300.
[9] A. Keshav, S. Chand, K.L. Wasewar, Extraction of propionic acid using different extractants (tri-n-butylphosphate, tri-n-octylamine, and aliquat 336), Ind. Eng. Chem. Res. 47 (2008) 6192–6196.
[10] A. Keshav, K.L. Wasewar, S. Chand, Reactive extraction of propionic acid using tri-n-octylamine, Chem. Eng. Comm. 197 (2010) 606–626.
[11] D. Datta, S. Kumar, B.V. Babu, Intensification of propionic acid separation from aqueous solution using reactive extraction with tri-n-butyl phosphate (TBP) dissolved in n-decane and 1-decanol. Proceedings of International Conference on Separation Processes (ICSP-2009), IT-BHU, Varanasi, October 20-22, (2009). 306-315.
[12] S. Kumar, B.V. Babu, Reactive extraction of propionic acid with Aliquat 336 dissolved in 1-decanol and n-dodecane. Proceedings of International Symposium & 61st Annual Session of IIChE in association with International Partners (CHEMCON-2008), Panjab University, Chandigarh, December 27-30, (2008).
[13] A. Flores-Morales, J. Albet, G. Kyuchoukov, G. Malmary, G. Molinier, Influence of extractant (TBP and TOA), diluent, and modifier on extraction equilibrium of mono-carboxylic acids. J. Chem. Eng. Data 48 (2003) 874-886.
[14] G.C. Ownubolu, B.V. Babu, New optimization techniques in engineering, Springer-Verilag, Heidelberg, Germany (2004).
[15] S. Kumar, D. Datta, B.V. Babu, Estimation of equilibrium parameters using differential evolution in reactive extraction of propionic acid by tri-n-butyl phosphate dissolved in *n*-decane & 1-decanol. Chem. Eng. Process. 50 (2011) 614-622.
[16] F.A. Poposka, J. Prochazka, R. Tomovska, K. Nikolovski, A. Grizo, Extraction of tartaric acid from aqueous solutions with tri-*iso*-octylamine (HOSTAREX A 324). Equilibrium and kinetics. Chem. Eng. Sci. 55 (2000) 1591-1604.

(498d) Accelerated Thermal Decomposition of Ammonia Borane by Carbon Dioxide

Junshe Zhang[1], Yu Zhao[2], Daniel L. Akins[2] and Jae W. Lee[1], (1)Department of Chemical Engineering, The City College of New York, New York, NY, (2)Department of Chemistry, The City College of New York, New York, NY

This talk presents the enhanced dehydrogenation of ammonia borane (AB) at 85 ^{o}C under CO2, an abundant and naturally occurring material. The presence of CO_2 significantly accelerates the thermal decomposition and it kinetics becomes faster as the CO_2 pressure increases from 1.70 to 3.67 bar. Around 4 bar of CO_2, 1.33 equivalents of hydrogen is released from AB in 1 hour. The accelerated AB thermolysis is attributed to CO_2 reacting with AB, or intermediates like diammoniate of diborane (DADB), or primary thermolysis products including linear dimers of aminoborane (LDAB), which accelerates AB thermolysis, and this reaction has been confirmed by ATR-FTIR, Raman, and ^{11}B and ^{13}C solid sate MAS-NMR measurements. The results open up a potential avenue to convert CO_2 to other useful chemicals under very mild conditions.

See more of this Session: Aspects of Hydrogen Storage Materials
See more of this Group/Topical: International Congress on Energy 2011

300 Year High Level Nuclear Waste

John L. Lyons, Edward J. Lahoda, Yasir Arafat, Fausto Franceschini
Westinghouse Electric Company LLC, Cranberry Township, PA, 16066, USA

Abstract

The nuclear industry is studying different options to improve the current nuclear fuel cycle. Most of these studies focus on either getting the most energy out of the system or optimizing the system economics. However, the waste that is produced by the current and many of the revised fuel cycles is not acceptable to the general public for a variety of reasons. This paper addresses what could be the major issue the public has, the >10,000 year lifetime of the toxicity of the radioactive waste. This paper focuses on the one approach to reducing and quantifying the radiohazard of the used nuclear fuel and explores the various quantifying options and justifies the best metric to determine the hazard involved with radioisotopes. The conclusion suggests a plausible regulation that would minimize the long-term hazard of used nuclear fuel and create a more benign waste.

Introduction

Although it is scientifically probable that one can keep high level waste isolated from the general population for a million years, people in general just do not believe that this task is credible. This credibility issue can severely impact the growth of nuclear as we face this nuclear renaissance. What the public will accept and view as credible is unknown, and is a job for social scientists to solve. Until it is known what the public will accept as credible, industry must formulate alternatives to the current scenario that will hopefully be more acceptable to the public.

The path put forward in this paper is centered on achieving an approach to high level nuclear waste that the public will accept as technically reasonable. One way to this is to look at what the public views as already acceptable. The EPA and chemical manufacturing companies are a good example of this. The EPA sets regulations on chemical companies for acceptance criteria on the disposal of wastes based on scientifically derived health standards for exposure, and the chemical companies must determine how to meet those criteria. What needs to be set for nuclear companies is a target criterion that the public could find to be reasonable and credible for the protection of human health. Nuclear wastes have an intrinsic advantage over chemical wastes in that they eventually decay and disappear while chemical wastes remain toxic for much longer. From there one could work backwards and formulate a series of possible alternative processes that will accomplish said criterion. After the possible alternatives are laid out, all the economics could be worked out and the cheapest, technically feasible process would be implemented.

Proposed Approach

The suggested criterion is that the disposed waste must have the same potential radiological impact to the public as the uranium ore that it originated from after 300 years. This waste, however, will occupy a much smaller volume hence it'll be much more concentrated than the U ore that it was derived from. Therefore, it will still be harmful in its concentrated form after 300 years. Nevertheless, in its concentrated form, this proposed waste will be contained within many engineered barriers to prevent exposure to the public. The concern for public safety arises if, by some unforeseen problem, the engineered barriers fail. If they were to fail, the radioactive matter that the public could potentially be exposed to must be no more harmful than that of U ore after 300 years. Whether the public will accept this proposed criteria is unknown, rather it is just a rational target that one may potentially get the majority to accept as realistic and protective of public health.

In order to determine whether a goal is feasible, one must first understand the target. In this case, the target is getting the final waste product to have the same potential radiological impact as that of U ore for the typical open cycle. Therefore, what is the potential radiological effect from U ore? U ore is made up of many different elements; on average U ores only contain 0.2% Uranium. However, a lot of uranium daughter isotopes are also radioactive and present in various quantities. The most abundant isotope in U ore is ^{238}U which has a half life of 4.46 billion years. The decay chain for ^{238}U is shown in Table 1.

The predominant decay mode of ^{238}U and its daughters is alpha decay then beta decay. Alpha decay is virtually harmless if kept outside the body. However, if ingested or inhaled, it is the most detrimental form of radioactivity. Beta decay is penetrating enough to cause some harm from outside the body, and will cause similar damage if ingested or inhaled.

Table 1– ^{238}U Decay Chain[1]

URANIUM DECAY CHAIN -- Main Branch Read from left to right. Arrows indicate decay.		
Uranium-238 ==> (half-life: 4.46 billion years) alpha decay	Thorium-234 ==> (half-life: 24.1 days) beta decay	Protactinium-234m ==> (half-life: 1.17 minutes) beta decay
Uranium-234 ==> (half-life: 245,000 years) alpha decay	Thorium-230 ==> (half-life: 75,400 years) alpha decay	Radium-226 ==> (half-life: 1,600 years) alpha decay
Radon-222 ==> (half-life: 3.82 days) alpha decay	Polonium-218 ==> (half-life: 3.11 minutes) alpha decay	Lead-214 ==> (half-life: 26.8 minutes) beta decay
Bismuth-214 ==> (half-life: 19.9 minutes) beta decay	Polonium-214 ==> (half-life: 163 microseconds) alpha decay	Lead-210 ==> (half-life: 22.3 years) beta decay
Bismuth-210 ==> (half-life: 5.01 days) beta decay	Polonium-210 ==> (half-life: 138 days) alpha decay	Lead-206 (stable)

^{235}U and its daughters are also a concern when discussing the radiological impact of U-ore. Its decay chain is, in many regards similar to that of ^{238}U. ^{235}U is an alpha emitter, and most of its daughters are as well. So, if one were to try and quantify what the radiological impact could be from U-ore, they would need to take into account all the daughter isotopes and a few assumptions. These assumptions would be:

1. How is a person exposed to the U-ore?
2. If ingested/inhaled how much of each isotope is ingested/inhaled?
3. Finally how long is this person exposed to these isotopes?

Once these assumptions have been made, a baseline value could be established for the dose associated with U-ore. However, this approach has one major flaw. There is no way to tie together the relationship

[1] Uranium: Its Uses and Hazards. *IEER.* [Online] July 2005. [Cited: March 22, 2011.] http://www.ieer.org/fctsheet/uranium.html.

with the U-ore consumed, and the amount of U-ore required to mine for a certain open cycle with x energy production. So for this problem, one could drop all previous assumptions and make a sole assumption. All the U-ore needed for the equivalent energy production from an open cycle, would be ingested by a theoretical person. The radiological impact of this ingestion could then be measure by a dose calculation. In this case a direct comparison could then be made between fuel cycles, on an energy production basis. Finally, since a baseline and comparison for different systems can be made, the original criterion of meeting 300 years can be further investigated for each proposed process.

Nuclear Fuel Cycle

One question that is often asked when looking at the nuclear cycle is will the Uranium supply ever run out? There isn't really a clear answer on this topic; there are strong cases to be made on both sides of the argument. Therefore, if one wants to make a switch from the traditional uranium cycle to another fertile/fissile cycle, other reasons must also be used.

An aspect to look at when considering any fertile/fissile combination is the waste that is generated when energy is produced. Each fertile/fissile combination will have its own unique waste that is produced; other factors such as how the reactor is set-up (geometry, burn-up, cycle length, etc.) will also change the waste that is produced. However, the latter will not be as drastic of a change as having a different fertile/fissile combination for energy producing reactors. In the typical $^{238}U/^{235}U$ cycle, isotopes of Np, Am, Cm and Pu are mainly produced through the transmutation of ^{238}U. Fission products are also formed from ^{235}U and some from ^{239}Pu which is bred from ^{238}U. A breakdown of a typical 238U/235U spent nuclear fuel from an LWR can be seen in Table 2.

Table 2– 33 GW/t PWR, 10 yr. Cooling*[2]

Uranium	Actinides	Long-Lived Fission Products (LLFPs)	Short-Lived Fission Products (SLFPs)	Stable Isotopes
955.4 kg	8.5 kg Pu	0.2 kg ^{129}I	1 kg ^{137}Cs	10.1 kg Lanthanides
	0.5 kg ^{237}Np	0.8 kg ^{99}Tc		
	0.6 kg Am	0.7 kg ^{93}Zr	0.7 kg ^{90}Sr	21.8 kg Other
	0.02 kg Cm	0.3 kg ^{135}Cs		

* 1 metric ton of SNF (Spent Nuclear Fuel)

In the United States, this spent fuel is currently placed into spent fuel pools or dry casks and held on site. Eventually something will need to be done with this spent fuel as reactor sites run out of capacity. As can be seen from Table 2, most of spent fuel (95.54%) is unused Uranium. There is also 0.85% Pu, some of which is fissile and could be reused. Of the remaining waste, 3.19% is stable. So, only 0.42% of the waste is both unsafe and unusable. There are three main options under consideration for dealing with SNF:

1) Once-through
2) Reprocess and reuse U and Pu in an LWR as a mixed oxide (MOX) fuel
3) Reprocess and reuse U, Pu, and MAs (Np, Am and Cm) in a Fast Reactor.

[2] **Salvatores, M.** Fuel Cycles for Sustainable Development and Waste Minimisation. *World Nuclear University.* [Online] July 2007. [Cited: March 23, 2011.] http://www.world-nuclear-university.org/html/summer_institute/2007/2007SI-lecture%20Materials/0725/0725_Massimo%20SALVATORES_1/0725-1-Massimo%20Salvatores.pdf.

1) Once-Through

In the once through scheme, the fuel is burned in a reactor one time, cooled for some predetermined amount of time and then placed in casks and buried in a final underground repository. This would be, by far, the cheapest route of disposal since no new technology would need to be investigated or applied. While extensive analysis of this approach has shown it to be a perfectly safe route, the public does not trust the credibility of this process due to the long time over which the fuel must be kept isolated from the public.

A modified open cycle has been proposed where the unused uranium is extracted from the waste and the rest of the waste would be placed in a cask and buried. The recycled U would then be enriched and mixed in with fresh fuel and reused in light water reactors (LWRs). This would increase the cost of the prototypical once through scheme, but could pose to be an incentive if it is found there will be a shortage of U resources. It would also reduce the total volume of what is buried in casks since roughly 95% of the SNF is U. This approach may have slightly more incentives than that of the typical open cycle; nonetheless, the MAs, LLFP, and Pu are all still being buried and the cost increase compared to immediate burial is substantial. These isotopes are the main contributors to the long term radiotoxicity of the waste because they remain radiologically dangerous between 100,000 and 1 million years after discharge and the size of the repository is also not reduced. Therefore, this approach is not favorably viewed by the public.

2) Reprocess and reuse U and Pu in an LWR

This process is also known as the MOX scheme. In this proposed process, SNF from an LWR would be reprocessed and U and Pu would be recycled back into a LWR to produce more energy. In this approach, the recycled U and Pu do create more energy. Unfortunately, they transmute into more MAs as well. These MAs would then be disposed of with the original MAs and hence from the same batch of U have even more MAs and greater radiological impact[3]. Or the MOX fuel would be set aside in anticipation of fast reactor development/use[4].

Ultimately, the fate of MOX fuel, if the final goal is to reduce the potential radiological impact of the waste, will be a fast reactor. So why should time, effort, and money be placed into the development of a technology that is simply a middle ground to the final solution. This leads to the third option in waste management.

3) Reprocess and reuse U in LWRs and the Pu and MAs in a Fast Reactor.

Reprocessing and reusing U in an LWR is very similar to that of the MOX scheme. The main difference is that the Pu and MAs are used in a FR (Fast Reactor), directly from the original LWR, which has a much harder spectrum than the LWR/MOX and many of the MAs are likely to fission to relatively short lived fission products. There will still be transmutation to some longer lived MAs but this transmutation will be more likely to lead to a fissionable isotope in the fast spectrum, than in the thermal. If there are more MAs fissioning, then there will be a more significant change in the lifetime of these isotopes than that of the open and MOX cycle.

A proposed alteration to the FR cycle is using ^{232}Th instead of ^{238}U as the fertile material. This would breed ^{233}U, which along with ^{232}Th is much further from the higher actinides, therefore, is much less likely to form new MAs. Transmutation products of concern, that are formed, are a few of the isotopes of Pa. These isotopes are also radioactive and have some potential radiological effects. It will

[3] **Arafat, Y., Lahoda, E.J. and Franceschini, F.** *Assessment of Technical Requirements for Partition Technologies to Reduce the Radiotoxicity of High Level Waste.* Westinghouse Electric Company, LLC. 2010.

[4] Mixed Oxide (MOX) Fuel. *World Nuclear Association.* [Online] March 21, 2009. [Cited: March 31, 2011.] http://www.world-nuclear.org/info/inf29.html.

need to be shown how different the waste will look from a radiological stand-point if the MAs are burned and replaced with Pa.

Fast Reactors also have some draw backs to their use. Although there has been a significant amount of research done, there has not been a significant amount of implementation of this technology. Also this technology typically uses a much more hazardous coolant than water (i.e. Sodium and Lead). Sodium reacts very violently with water and air (because of the water vapor in air), and Lead is toxic to humans. However, there is a significant part of the scientific community that believes the benefits out-weigh the negatives, and that the negatives can be safely implemented so that they are not a significant technical problem. It remains to be seen if the public can also be convinced that the safety problems are also insignificant.

Radiological Impact

The discussion on how one should calculate the radiological impact of spent nuclear fuel is a contentious one. On one side you have the argument that repositories, casks, and other engineered barriers are designed to contain all the nuclear waste and the probability of these contained isotopes reaching the biosphere and endangering society is minimal. Even if these isotopes were to leak from their containment, the ingestion probability remains very low. This is because of the slow moving ground water at depth, the immobility of the isotope complexes that are present, and the distance of a repository to the ground water used by people[5]. Though these claims are sound scientifically, there are others that also believe that one must also look at not only the natural progression of events, but also an accelerated version, where-in people tap into the potentially contaminated ground water/soil prematurely.

In the first instance studies have shown that, with the natural progression of radioactive nuclides reaching people from an underground repository, these nuclides would be no more harmful than that of natural uranium ore deposits[5]. Many studies have been done by various groups, such as the ICRP (International Commission on Radiological Protection) and several government agencies that have focused on what the effect is on the body if people were to come in contact with the various radioactive isotopes. The later studies can be applied when one considers a person interfering with the natural progression of events and tapping into the radiation hazard prematurely. Regardless of how one comes in contact with a radioactive isotope, equation 1 is widely accepted as the proper way to calculate the affect that isotope has on the body.

Dose (Sv or rem) = Activity (Bq or Ci) x Dose Coefficient [1]

This equation may change slightly depending on what type of dose is of interest. For example in the ORNL document, <u>External Exposure to Radionuclides in Air, Water, and Soil</u> presented to the EPA, their dose coefficients have units of (Sv m^x Bq^{-1} s^{-1}) where x is either 2 or 3 depending on what medium the radionuclides are found in[6]. For example, if the radionuclides are spread out across the ground surface one would use the m^2 dimension. But if the radionuclides are mixed in with the soil, are immersed in water, or are suspended in air, the m^3 dimension would be used. As can be seen, this report is focused on the external dose associated with a radionuclide. Since external dose is exclusively governed by time, distance, and shielding, the activity used must be in units of (Bq s m^{-x}). This activity

---

[5] **Benedict, M., Pigford, T. H. and Levi, H. W.** *Nuclear Chemical Engineering.* Second Edition. Boston : McGraw-Hill, Inc., 1981. 0-07-004531-3.

[6] **Eckerman, Keith F. and Ryman, Jeffrey C.** *External Exposure to Radionuclides in Air, Water, and Soil.* Oak Ridge : Oak Ridge National Laboratory, 1993.

unit would take into account how much of the radionuclide is present, how long the individual was exposed, and the concentration of said radionuclide in the environment. This activity multiplied by the corresponding dose coefficient will render the dose based solely on the assumption that the radionuclides remain outside the body.

In contrast, in publication 72 from the ICRP, the main interest is on the internal effects of radioactive isotopes. When considering dose from nuclides within the body, the main factors in controlling said dose are: how long the isotope is retained within the body, what type of radiation the isotope is emitting, and how much of this isotope is ingested/inhaled. The ICRP incorporates all the retention information and what type of radiation is being emitted into the dose coefficients. In other words, the dose coefficients carry units of ($Sv\ Bq^{-1}$). So the required activity is solely Bq, or through simple conversion Ci, which represents how much of the nuclide was consumed. The resulting product of the activity and dose coefficient will be the dose from the retained radionuclide based on the assumed retention times.

The ingestion of radionuclides would likely be the best approach since this is the most detrimental to public safety and the most likely mode of exposure. Although referred to in many ways (dose, ingestion dose, radiotoxicity, etc.), all have the same base units. The one exception is radiotoxicity which is sometimes seen with the units of m^3 water. This representation of radiotoxicity is simply a dilution based on the allowed regulated exposure per person per year. This representation can easily be reverted back to Sv or Rad, to reflect an absorbed dose.

In order to compare multiple cycles and determine whether a fuel cycle meets the designated criteria, one must tie the total dose or radiotoxicity to that of the total energy produce by that fuel cycle. Then one can find the total amount of U-ore that would need to be mined in order to meet that energy production in a typical open cycle, and find the total radiotoxicity of that U-ore. The U-ore radiotoxicity will give a baseline and the total radiotoxicity at different time-points from the desired fuel cycle can be plotted together to determine whether the goal meets the 300-year criteria.

An Approach to the Nuclear Cycle

A LWR-FR cycle can be used to achieve the 300 year goal. In this proposed cycle, LWR spent fuel is cooled and reprocessed. After reprocessing, Pu and Np are mixed with ^{232}Th and driver fuel rods/assemblies are produced to drive the nuclear reaction in a FR. The MAs from the LWR SNF are put into TRISO beads and these beads are put into a SiC matrix[7]. These rods are then formed into target assemblies for the FR. The driver fuel and targets are then burned for a given cycle length in a FR. Some of the driver fuel is then removed, cooled, and reprocessed to recover any unused ^{232}Th, Np and Pu. Also recovered is any ^{233}U that has been bred. The fission products are sent to HLW and any Pa/MAs from the transmutation of the Np and Pu are sent to TRISO manufacturing to be placed within beads and put back into the FR. The fission products are sent to HLW because most will undergo 10 half-lives before or around the 300 year criteria limit. After the LWR SNF cools and goes to the reprocessing center, the rest of the process including reprocessing will be contained within a central area to insure security and safety (see Figure 1).

[7] **Jo, C.K., Kim, Y. and Venneri, F.** *DB TRISO Fuel in LWRs.* [pdf] Oakridge, Tennessee : s.n., September 30, 2010.

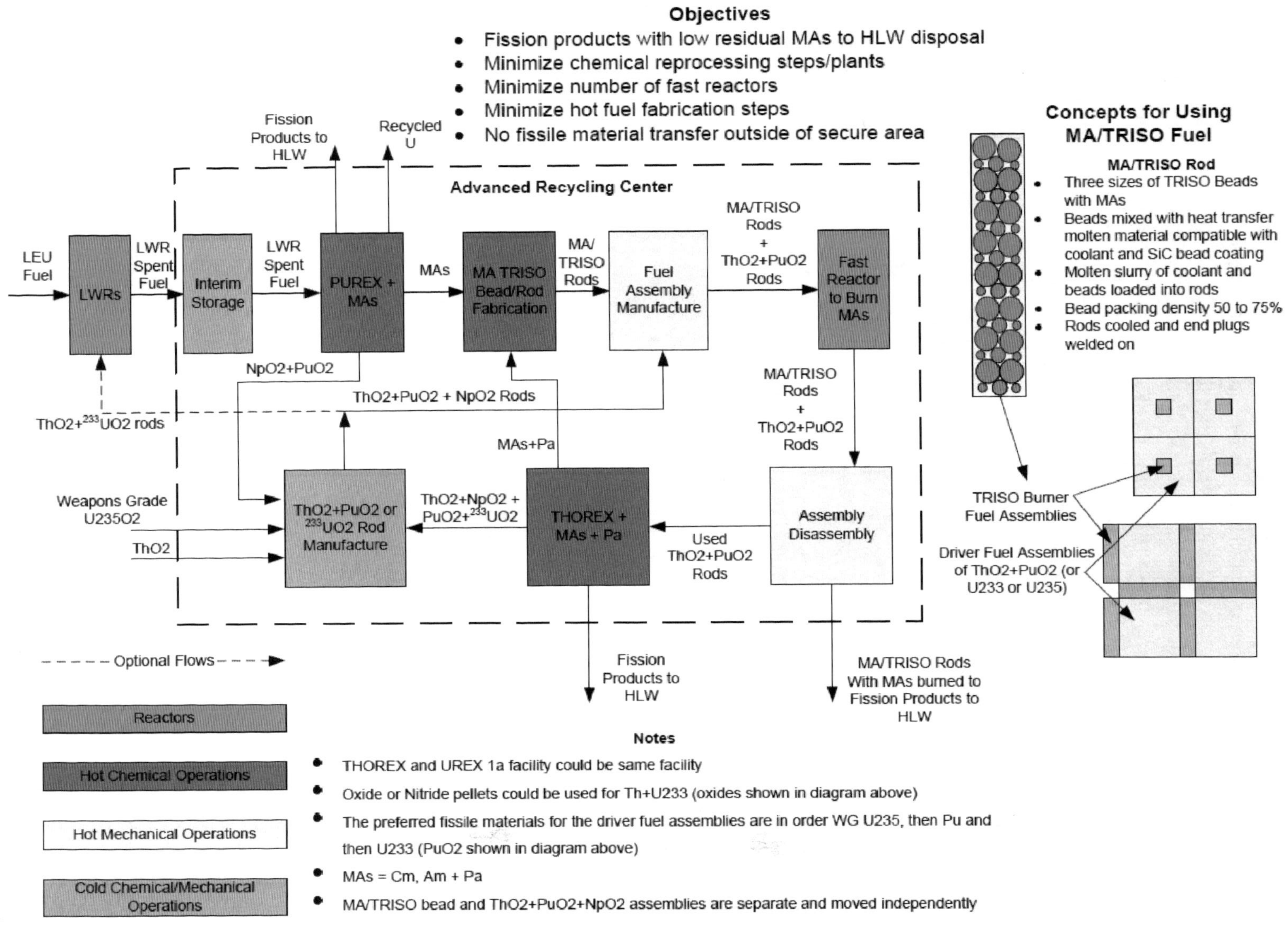

Figure 1 – Proposed Nuclear Cycle Approach

Conclusion

The key to public acceptance of nuclear waste is to present either a reasonable explanation as to why direct disposal of spent fuel is acceptable or to propose a reasonable alternative. As an alternative, the potential radiological impact on the public can be reduced by partitioning and transmuting a majority of the long-lived minor actinides and plutonium such that the residual high level waste has a lifetime of ~300 years at which time it will be no more radiotoxic than the ore from which it was derived. Since the highest potential risk for the public comes in when considering someone drinking contaminated water or eating contaminated food, the only dose that should be considered is that of ingestion. Although typically this would mean that when calculating dose from ingested nuclides, either assumptions need to be made as to how this isotope got within reach of an individual, or a transport model would need to be developed to show through calculation how the individual accessed the isotope. However, it is proposed that a single assumption be made; a theoretical person accesses and ingests the HLW from a specific process, and the equivalent U-ore that is associated with that process. The effects from HLW ingestion and U-ore ingestion can then be compared and evaluated, without the direct need for a transport model. This is a sound assumption because in their disposed form the main chemicals in question (TRUs) behave very similarly to one another. So, the transport of each nuclide can be viewed as equal to that of the other TRU nuclides.

The technical feasibility, of various options including the LWR-FR cycle mentioned above, is still being analyzed to determine the technical requirements and costs.

(450b) Development of Zr-89 PET Radiopharmaceuticals for Cancer Imaging

Jason S. Lewis, Radiology, Memorial Sloan-Kettering Cancer Center, New York, NY

The use of Positron Emission Tomography (PET) for cancer imaging is a well-established and widely used molecular imaging modality both in clinical and research settings. PET offers the ability to quantitatively measure biological and receptor-based processes using a wide spectrum of specifically designed radiopharmaceuticals. The use of PET is expanding and the inclusion of longer-lived radiometal positron-emitters is broadening the application and appeal of this imaging modality.

Zr-89 has emerged as a promising radionuclide for use in immunoPET. Zirconium-89 has a number of distinct advantages which make it ideal for immunoPET: (i) the radioactive half-life of 78.41 h matches closely the extend times required for optimum biodistribution of intact mAbs, (ii) the positron yield of 22.7% is comparable to that of Cu-64, Y-86 and I-124 which improves counting statistics in PET imaging, (iii) zirconium and its ions are generally inert to biological systems and have no known biological role or function, (iv) cyclotron production of Zr-89 via the (p,n) transmutation reaction using a 100% naturally abundant Y-89 solid target is highly efficient and cost effective, and (v) high purity and high specific-activity Zr-89 is now available in various chemical forms which are suitable for radiolabeling mAbs.

This presentation will review the current state-of-the-art in non-standard PET nuclide application with an emphasis on the use of radiometals (e.g., Zr-89) with peptide and antibody constructs. In addition, since the effective use of a radiometal nuclide often relies on their attachment to the targeting probe via a bifunctional chelator, this talk will focus on novel strategies for using "click" chemistry methodology for attachment of the radiometals to active biomolecules.

See more of this Session: Engineering Applications of Radioisotopes for Industry and Medicine
See more of this Group/Topical: Nuclear Engineering Division

(213b) Behavior of Mercury During Chemical Processing In the Defense Waste Processing Facility

John R. Zamecnik, David C Koopman and Michael E Stone, Savannah River National Laboratory, Aiken, SC

The Defense Waste Processing Facility (DWPF) at the Savannah River Site vitrifies High Level Waste (HLW) for disposal. DWPF has been processing radioactive waste since March 1996 to produce over 3,000 canisters filled with HLW glass. Sludge waste consists of insoluble metal hydroxides (primarily iron, aluminum, calcium, magnesium, manganese, and uranium) and soluble sodium salts (carbonate, hydroxide, nitrite, nitrate, and sulfate).

The pretreatment process in the Chemical Processing Cell (CPC) Sludge Receipt and Adjustment Tank (SRAT) adds nitric and formic acids to the sludge to lower pH, destroy nitrite and carbonate, and chemically reduce mercury and manganese. Acidification of the sludge also allows higher solids concentrations to be processed by lowering the yield stress of the sludge. During these processes, hydrogen can be produced by noble metal catalyzed decomposition of excess formic acid. The ratio of nitric to formic acid is adjusted to target a glass pool with 20% of the iron having a valence of Fe^{2+}.

The feed to the SRAT contains mercury as mercuric oxide (HgO). The reduction by formic acid converts the HgO to elemental mercury which is subsequently stripped off by boiling. The maximum concentration of mercury in allowed in the product is about 0.8 wt% of the total solids to reduce deposition of mercury compounds in the melter offgas system. The theoretical stripping ratio, based on vapor pressures, is 1g mercury evaporated per 250g water evaporated; early studies showed that the actual ratio was about 1:750. Recent experience in DWPF has shown that the ratio is often much greater than 1:750 resulting in extremely long stripping times. Also, recovery of mercury in the offgas system in a decanter has been poor.

Laboratory studies were performed to improve the understanding of the mechanisms of mercury reduction, stripping, and recovery. The effects of other redox species, chlorides, mercury drop size, and slurry concentration on reduction and stripping were examined. Recovery of mercury was studied to determine the effect of non-condensable gases, and other species condensed with mercury that may chemically react or inhibit condensation and coalescence.

See more of this Session: Contaminant Transport and Site Remediation
See more of this Group/Topical: Environmental Division

Simulation of Reactive Extraction Columns Based on Lab-scale Experiments with Single Drops.

Murat Kalem, Florian Buchbender, Andreas Pfennig

RWTH Aachen University, AVT – Thermal Process Engineering, Germany

www.avt.rwth-aachen.de

Reactive extraction is not only used to refine noble metals or to enrich nuclear fuels but also to separate components of low concentration from a liquid phase. Due to superimposed reaction this is more efficient than by conventional solvent extraction. Compared to solvent extraction, the interactions in reactive extraction are more complex due to various reactions that may be taking place either in the bulk phases or at the interface parallel to physical mass transfer. For the reliable and efficient design of such processes, quantitative knowledge of mass transfer and reactions and their effect on the hydrodynamics in an extraction column is crucial.

A new method based on simulation with drop population balances and lab-scale single-drop experiments allows for understanding these effects in detail. The single-drop experiments are used to obtain system-specific parameters for models accounting for mass transfer and drop sedimentation. At AVT – Thermal Process Engineering, a drop population balance tool called ReDrop, originally developed and extensively verified for solvent extraction, has now been extended for reactive systems. Single-drop experiments with the standard test system for reactive extraction zinc + D2EHPA have been conducted. It was shown that, in contrast to physical extraction systems, for this test system, the interfacial reaction increases the drop's sedimentation velocity. This is because D2EHPA, acting as a surface active agent that makes the interface rigid, is consequently removed from the surface by the interfacial reaction with zinc. As a consequence of this increased mobility of the interface, mass transfer is enhanced. To model this phenomena, existing mass-transfer models were extended and incorporated into ReDrop. Multi-component reactive equilibria were incorporated and G^E models were used to account for the non-idealities of both phases. For different operating conditions of a pulsed sieve-tray column, simulation results like Sauter mean diameter, hold-up and separation efficiency were then compared to data from pilot-plant experiments.

In the presentation, the results of the single-drop experiments for the reactive test system will be presented and discussed. It will be shown that accurate simulation of reactive extraction columns is possible employing single drop experiments and ReDrop.

A Future Energy Supply Based on Liquid Organic Hydrogen Carriers (LOHC)

<u>Daniel Teichmann</u>, Wolfgang Arlt, Peter Wasserscheid
University of Erlangen-Nuremberg, Germany

Introduction

Our current energy system is mainly based on the availability of fossil fuels that can be extracted from the ground to a comparably cheap price and that are also excellent energy carriers in terms of storage density and manageability. The ongoing transition towards renewable sources is a big challenge for our energy system, mainly due to the highly intermittent character of many renewable energies like wind or solar.

While hydrogen is often considered as a suitable energy vector in a future sustainable energy system [1,2], the weight and volume-efficient storage of sufficient amounts of hydrogen is a difficult task due to its low density. A comprehensive hydrogen infrastructure does not exist yet and makes high investments necessary.

Liquid Organic Hydrogen Carriers (LOHC)

This contribution describes a concept for the establishment of a competitive energy distribution network based on Liquid Organic Hydrogen Carrier (LOHC) compounds [3]. These compounds are characterized by the fact that they can be loaded and unloaded with considerable amounts of hydrogen in a cyclic process. In contrast to other hydrogen storage methods (like compressed gaseous or cryogenic hydrogen) LOHCs can be safely stored under ambient conditions in big quantities. As they have many physico-chemical similarities to Diesel, LOHCs can make use of the existing energy infrastructure (e.g. storage tanks or road tankers).

Many different substances are discussed in the literature as potential LOHC compounds [4,5]. In this contribution we concentrate on heterocyclic aromatic hydrocarbons such as N-ethylcarbazole (Fig. 1) which were first considered by the company AirProducts [6]. The fully hydrogenated substance perhydro-N-ethylcarbazole offers a hydrogen capacity of up to 5.8 wt%. The hydrogenation reaction for this particular LOHC system is typically carried out at 130–160 °C with 70 bars pressure using supported noble metal catalysts like ruthenium on aluminum oxide. The dehydrogenation reaction proceeds usually in the presence of supported Pt or Pd catalysts. The reaction is typically carried out at 200–230 °C and pressures slightly above ambient conditions [7-9].

Energy storage and mobility applications

In times of high loads from intermittent sources like wind or solar energies, excess energy can be used to locally produce hydrogen by water electrolysis, which is efficiently stored in the form of LOHC. The so-produced energy carrier can be stored at the spot of production and re-converted to electricity in times of high demand.

Alternatively it can be easily transported to be used in various off-site applications like mobility (as discussed in the following), heating, local hydrogen production, etc.

Compared to other energy storage technologies like pumped hydroelectric or compressed air energy storage [10], LOHCs show a high energy storage density, thus enabling a safe and loss-free storage of renewable energy with only limited land use. These reasons make LOHCs a promising technology especially for seasonal energy storage where large amounts of energy have to be stored over extended periods of time.

Another promising field of application for LOHCs is mobility [11]. To allow cruising ranges of 500km, about 5 kg of on-board hydrogen are necessary which correspond to about 100 liters of carrier compound. The hydrogen which is produced from the LOHC on-board in a catalytic converter can either be used in a combustion engine or a fuel cell (see Fig. 2). At the gas station the unloaded carrier compound is returned in exchange to fresh material and can be reloaded with hydrogen for further use.

Conclusion

The proposed concept links the technical challenge of storing temporary and local energy over-production from regenerative sources with the vision of a sustainable, hydrogen-based mobility. Due to their beneficial handling and storage properties we consider LOHCs as an attractive way to provide wind and solar energy for mobility applications.

Figures

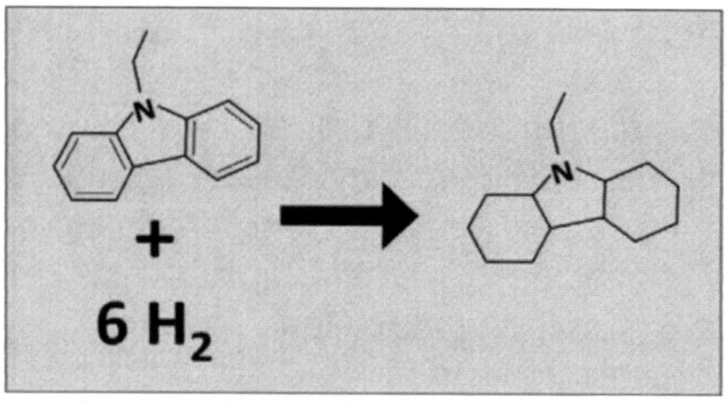

Figure 1: Chemical energy storage via hydrogenation of a LOHC compound [3]

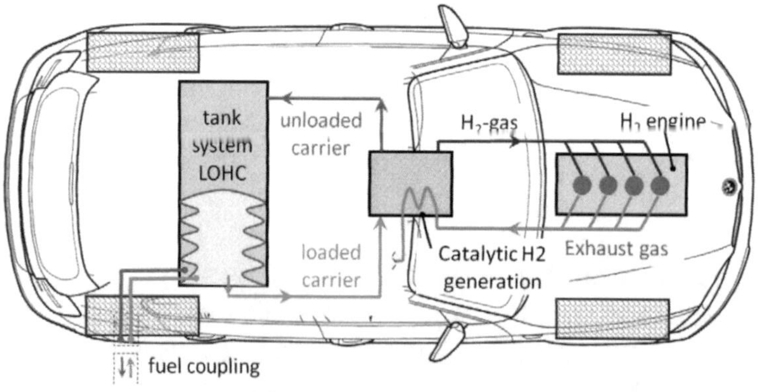

Figure 2: Concept of a hydrogen car with liquid organic hydrides [3]

References

1. Sartbaeva, V. L. Kuznetsov, S. A. Wells and P. P. Edwards, Energy Environ. Sci., 2008, 1, 79–85.
2. U.S. Department of Energy, Hydrogen Posture Plan, http://www.hydrogen.energy.gov.
3. D. Teichmann, W. Arlt, P. Wasserscheid, R. Freymann, Energy Environ. Sci., 2011, 4 (8), 2767 – 2773
4. G. W. H. Scherer, E. Newson, A. Wokaun, Int. J. Hydrogen Energy, 1999, 24, 1157–1169.
5. R. H. Crabtree, Energy Environ. Sci., 2008, 1, 134–138.
6. Pez et al., World Pat., WO 2005/000457A2, 2005.
7. M. Sobota et al., Chem.–Eur. J., DOI: 10.1002/chem.201101311.
8. F. Sotoodeh and K. J. Smith, Ind. Eng. Chem. Res., 2010, 49, 1018–1026.
9. K. M. Eblagon, D. Rentsch, O. Friedrichs, A. Remhof, A. Züttel, A. J. Ramirez-Cuesta and S. C. Tsang, Int. J. Hydrogen Energy, 2010, 35, 11609–11621.
10. Verband der Elektrotechnik, Elektronik, Informationstechnik (VDE), Energiespeicher in Stromversorgungssystemen mit hohem Anteil Erneuerbarer Energieträger, Frankfurt, 2009.
11. J. von Wild, R. Freymann and M. Zenner, Potentiale von alternativen Wasserstoffspeicherungstechnologien, VDI-Berichte: Innovative Fahrzeugantriebe, 2008, vol. 2030, pp. 273–298.

(343c) Asset Revitalization for Cleaned-up Department of Energy Sites

Sharon Robinson, Oak Ridge National Laboratory, Oak Ridge, TN

For decades, Department of Energy (DOE) sites and their surrounding communities have played a vital role in supporting the nation's nuclear weapons defense mission. Thanks to investments through the American Recovery and Reinvestment Act, many sites that were critical to supporting national defense are poised to build a new future. DOE footprint reduction and environmental cleanup efforts have created opportunities to transition existing Federal properties and other assets to future beneficial uses. This paper highlights some of the activities underway and potential opportunities to reutilize DOE sites assets for beneficial purposes, such as clean energy development, environmental cleanup sustainability projects, open space or other uses.

See more of this Session: Wilson Award Plenary Session on Nuclear Chemical Engineering (Invited Papers)
See more of this Group/Topical: Nuclear Engineering Division

Flowsheet Alternative for Chemical Processing of Savannah River Site High Level Radioactive Waste to Minimize Hydrogen Generation

Dan P. Lambert, Savannah River National Laboratory, Aiken, SC, USA
Bradley R. Pickenheim Savannah River National Laboratory, Aiken, SC, USA
Michael .E. Stone, Savannah River National Laboratory, Aiken, SC, USA
J. David Newell, Savannah River National Laboratory, Aiken, SC, USA
David R. Best, Savannah River National Laboratory, Aiken, SC, USA

EXECUTIVE SUMMARY

Flowsheet testing was performed to develop the nitric-glycolic-formic acid flowsheet (referred to as the glycolic-formic flowsheet throughout the rest of the report) as an alternative to the nitric/formic flowsheet currently being processed at the DWPF. This new flowsheet has shown that mercury can be removed in the Sludge Receipt and Adjustment Tank (SRAT) with minimal hydrogen generation. All processing objectives were also met, including greatly reducing the Slurry Mix Evaporator (SME) product yield stress as compared to the baseline nitric/formic flowsheet. Forty-six runs were performed in total, including the baseline run and the melter feed preparation runs. Significant results are summarized below:

Constraint	Limit	Baseline Flowsheet GF1	Glycolic-formic Flowsheet GF3
SRAT hydrogen, lb/hr	<0.65	1.62	0.03
SME hydrogen, lb/hr	<0.23	0.0072	0.0017
SRAT carbon dioxide, lb/hr	NA	375	200
SRAT nitrous oxide, lb/hr	NA	0.75	1.93
SRAT product Hg, wt %.	0.8	0.66	0.56
SRAT product nitrite, mg/kg	<1000	<100	<100
SRAT product down yield stress, Pa	1.5 to 5	33.1	1.6
SRAT product down consistency, cP	5 to 12	22.8	7.1
SME product down yield stress, Pa	2.5 to 15	223	7.2
SME product down consistency, cP	10 to 40	289	24.1
Glass REDOX $Fe^{+2}/\Sigma Fe$	0.1-0.33	0.00	0.22
SME product total solids, wt %	>45	46.28	44.66
Minimal foaming	NA	No	Yes

The baseline nitric/formic flowsheet run, using the SB6 simulant produced by Harrell was extremely difficult to process successfully under existing DWPF acceptance criteria with this simulant at the HM levels of noble metals. While nitrite was destroyed and mercury was removed to near the DWPF limit, the rheology of the SRAT and SME products were well above design basis and hydrogen generation far exceeded the DWPF SRAT limit. In addition, mixing during the SME cycle was very poor. In this sense, the nitric/glycolic/formic acid flowsheet represents a significant upgrade over the current flowsheet. Mercury was successfully removed with almost no hydrogen generation and the SRAT and SME products yield stresses were within process limits or previously processed ranges.

The glycolic-formic flowsheet has a very wide processing window. Testing was completed from 100% to 200% of acid stoichiometry and using a glycolic-formic mixture from 40% to 100% glycolic acid. The testing met all processing requirements throughout these processing windows. This should allow processing at an acid stoichiometry of 100% and a glycolic-formic mixture of 80% glycolic acid with minimal hydrogen generation. It should also allow processing endpoints in the SRAT and SME at significantly higher total solids content and may be effective at acid stoichiometries below 100%, although no testing was performed below 100% acid stoichiometry.

There are several issues related to the development of the glycolic-formic flowsheet. First, the measurement of anions using the new glycolate anion procedure likely needs to be optimized to improve the accuracy of the anions important to DWPF processing and REDOX prediction. Second, the existing REDOX equation with an added term for glycolate did not accurately predict the glass REDOX for the glycolic-formic flowsheet. Improvement of the anion measurement or modification of the REDOX methodology or equation may be necessary to improve the REDOX prediction. Last, the glycolic-formic flowsheet dissolves a number of metals, including iron. This leads to a thinner slurries but also dissolves up a portion of the iron, which is currently used for criticality control.

It is recommended that DWPF continue to support development of the glycolic-formic flowsheet. This flowsheet meets or outperforms the baseline flowsheet in off-gas generation, mercury removal, product rheology and general ease of processing. Additional testing is in progress to demonstrate the effectiveness of the nitric-glycolic-formic flowsheet in processing a wide sludge processing window using the matrix sludge simulants.

TABLE OF CONTENTS

LIST OF TABLES

LIST OF FIGURES

LIST OF ABBREVIATIONS

AD	Analytical Development
ARP	Actinide Removal Process
CPC	Chemical Process Cell
GC	Gas Chromatograph
FAVC	Formic Acid Vent Condenser
IC	Ion Chromatography
ICP-AES	Inductively Coupled Plasma-Atomic Emission Spectroscopy
MCU	Modular Caustic Side Solvent Extraction Unit
MWWT	Mercury Water Wash Tank
PSAL	Process Science Analytical Laboratory
REDOX	Reduction/Oxidation
SME	Slurry Mix Evaporator
SMECT	Slurry Mix Evaporator Condensate Tank
SRAT	Sludge Receipt and Adjustment Tank
SRNL	Savannah River National Laboratory
SRR	Savannah River Remediation
SVOA	Semi-Volatile Organics Analysis
TIC	Total Inorganic Carbon Analysis
TOC	Total Organic Carbon Analysis
TT&QAP	Task Technical and Quality Assurance Plan
TTR	Technical Task Request
VOA	Volatile/semi-volatile Organics Analysis
VSL	Vitreous State Laboratory

Introduction

Savannah River Remediation (SRR) is evaluating changes to its current Defense Waste Processing Facility (DWPF) flowsheet to improve processing cycle times that will enable the facility to support higher canister production while maximizing waste loading after installation of the bubblers into the melter. Due to the significant maintenance required for the DWPF Gas Chromatographs (GC) and the potential for production of flammable quantities of hydrogen, reducing the amount of formic acid used in the Chemical Process Cell (CPC) is one of the options being considered. Earlier work at Savannah River National Laboratory has shown that replacing formic acid in the existing nitric/formic acid flowsheet with an 80:20 molar blend of glycolic and formic acids has the potential to remove mercury in the SRAT without any significant catalytic hydrogen generation.

Evaluation of Alternative Reductants

At the request of SRR, a study was completed evaluating 19 reductants to replace formic acid in CPC Processing[1]. In the DWPF CPC, formic acid is both an acid and a reductant. Six reductants that are not acids were evaluated in addition to thirteen reducing acids. The two best alternatives from this testing, glycolic and acrylic, are listed in Table 1-1. All. The primary drawback to acrylic acid, the only acid tested with a carbon double bond, is the potential for polymerization in the storage tank, requiring controls to be in place to allow safe storage. Since this initial study, all testing has been completed with glycolic acid, which is also both an acid and a reductant. Note that for every mole of glycolic acid added, two moles of carbon are added as glycolic acid is a two carbon organic acid. Glycolic mixed with formic acid was recommended to ensure mercury would be reduced even if glycolic acid was not effective in reducing mercury.[2]

Table **0-1**. Acid Properties

Reductant	Formula	Acid pKa	Solubility 25 [illegible]C g/100 ml	Molarity	Typical Acid Concentration	Carbon Oxidation State	Structure
Formic Acid	CH_2O_2	3.751	Miscible	23.6	90 wt %	2	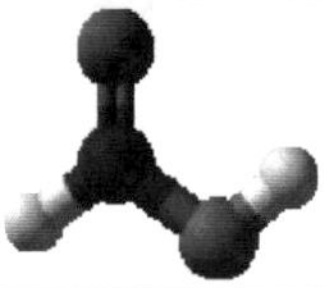
Acrylic Acid	$C_3H_4O_2$	4.35	Miscible	17.5	100 wt %	0	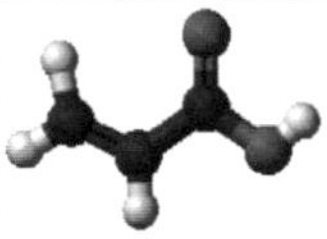
Glycolic Acid	$C_2H_4O_3$	3.831	80	11.83	71 wt%	1	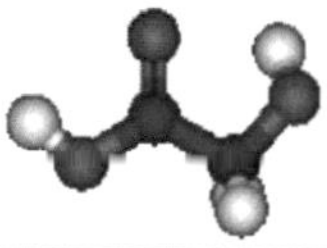

[1] Pickenheim, B.R., M.E. Stone, D.K. Peeler, Selection and Preliminary Evaluation of Alternative Reductants for SRAT Processing, SRNL-STI-2009-00120, Savannah River National Laboratory, Aiken, SC, June 2009.

[2] Pickenheim, B.R., M.E. Stone, SRAT Alternative Reductant Feasibility Assessment – Phase I, SRNL-STI-2009-00120, Savannah River National Laboratory, Aiken, SC, February 2009.

Development of Glycolic-Formic Flowsheet

The objective of the testing detailed in this document is to summarize the data collected in developing the glycolic-formic acid flowsheet as requested by DWPF.[3,4,5] This work was performed under the guidance of Task Technical and Quality Assurance Plans (TT&QAP).[6,7,8]

The sludge simulant required to complete this testing was procured from an off-site vendor. This simulant is being used to support all of the alternative flowsheet testing for the downselect process. The details regarding the simulant preparation and analysis have been documented separately.[9] The simulant was based on Sludge Batch 6 with HM basis noble metal concentrations (Table 1-2).

Table **0-2**. Mercury and noble metal concentrations

Element	Concentration, wt % (solids basis)
Hg	3.263
Ag	0.014
Pd	0.079
Rh	0.038
Ru	0.217

A total of twenty-two CPC simulations including SRAT and some SME cycles were performed. The first four tests were a baseline nitric-formic flowsheet, a baseline nitric-glycolic-formic flowsheet, a run without mercury and a run with Actinide Removal Process (ARP) and Modular Caustic Side Solvent Extraction (MCU) streams added. The second set of four simulations included tests at varying acid stoichiometries to define the acid processing window and one test without any formic acid (nitric/glycolic acid only) to determine the effectiveness of glycolic acid as a reductant. No SME cycle

---

3 Holtzscheiter, E. W. Technical Task Request – Slurry Receipt and Adjustment Tank (SRAT) Alternative Reductant Assessment, HLW-DWPF-TTR-2008-0039, Washington Savannah River Company, Aiken, SC, 2008.

4 Holtzscheiter, E. W. Technical Task Request – Perform Glycolic-Formic Acid Flowsheet Development, Definition, and Demonstration, HLW-DWPF-TTR-2010-0003, Savannah River Remediation, Aiken, SC, December 2009.

5 Holtzscheiter, E. W. Technical Task Request – Perform Glycolic-Formic Acid Flowsheet Development, Definition, and Demonstration, HLW-DWPF-TTR-2010-0003, Revision 1, Savannah River Remediation, Aiken, SC, September 2010.

6 Pickenheim, B.R., Task Technical and Quality Assurance Plan for Glycolic-Formic Acid Flowsheet Development, Definition and Demonstration Tasks 1-3, SRNL-RP-2010-00105, Savannah River National Laboratory, Aiken, SC, Add Date.

7 Pickenheim, B.R., Task Technical and Quality Assurance Plan for Glycolic-Formic Acid Flowsheet Development, Definition and Demonstration Tasks 4-5, SRNL-RP-2009-00484, Savannah River National Laboratory, Aiken, SC, April 2010.

8 Pickenheim, B.R., Choi, A.S., Task Technical and Quality Assurance Plan for Glycolic-Formic Acid Flowsheet Development, Definition and Demonstration Tasks 4-7A, SRNL-RP-2009-00484, Revision 2, Savannah River National Laboratory, Aiken, SC, October 2010.

9 Fernandez, A.I., SB6-H Simulant Development: Evaluation Memo, SRNL-L3100-2010-00159, Savannah River National Laboratory, Aiken, SC, August 2010.

was performed on the glycolic acid only flowsheet simulation[10]. Four tests were completed to produce products with a REDOX of 0-0.3 to demonstrate REDOX control for this flowsheet. Four tests were completed at glycolic-formic acid blends of 40-70% glycolic acid to determine the iron solubility at varying glycolic acid blends. Lastly, two runs were completed without mercury and noble metals to determine the conditions for producing melter feed with the baseline and glycolic-formic flowsheets as described in Section 1.3. A more complete summary of the runs is included in Table 2-1.

Total boiling time in the SRAT cycles were calculated to remove mercury to 0.60 wt% in the total solids at a stripping rate of 750 lb steam/lb Hg at the scaled maximum DWPF design rate of 5000 lb/hr steam. Process samples were taken during the runs to monitor mercury concentration with time. Off-gas data were collected to monitor hydrogen as well as CO_2 and N_2O generation.

The amount of acid used in each simulation was calculated using the Koopman minimum acid equation.[11] A stoichiometric factor of 125% was used for the baseline runs and factors varying from 100% to 200% were used for the acid window testing. The Hsu equivalents for these runs are 135% stoichiometry for the baseline and 108% to 215% for the acid window runs. REDOX was targeted at 0.2 $Fe^{+2}/\Sigma Fe$ using a modified REDOX equation with a term for glycolate ion included.

Preparation of Melter Feed for Testing of Baseline, Glycolic-Formic and Sugar Flowsheets

A total of twenty-four shortened SRAT cycles were performed to produce sufficient melter feed to test the baseline, glycolic-formic and sugar flowsheets in the VSL DM10 melter. Approximately 100 kg (25 gallons) of SRAT product was produced for each flowsheet. No noble metals or mercury were added to the sludge in these experiments. The same simulant was used for each of these runs.

---

[10] Pickenheim, B.R., M.E. Stone, J.D. Newell, Glycolic-Formic Acid Flowsheet Development, SRNL-STI-2010-00523, Rev 0, Savannah River National Laboratory, Aiken, SC, November 2010.

[11] Koopman, D.C., A.I. Fernandez and B.R. Pickenheim, Preliminary Evaluations of Two Proposed Stoichiometric Acid Equations, SRNL-L3100-2009-00146, Savannah River National Laboratory, Aiken, SC, 2009

Experimental Procedure

The experimental apparatus used in these experiments is typical for DWPF SRAT/SME testing. The first twenty-two experiments were performed in 4-L kettles, while the twenty-four melter feed runs were performed in 22-L kettles. In experiments with noble metals and mercury, test equipment included a gas chromatograph to measure offgas composition, an ammonia scrubber, and a pH meter. In experiments without noble metals and mercury, the use of this equipment was omitted. The one exception to this is that a pH meter was used in the first sugar flowsheet melter feed preparation run. In all runs except the sugar flowsheet runs, the SRNL acid calculation spreadsheet used the Koopman Equation to determine acid addition quantities and dewater targets. In the sugar flowsheet runs, the run parameters were defined by the Vitreous State Laboratory (VSL)[12].

CPC Simulation Details

The first twenty-two glycolic-formic flowsheet tests were performed at the ACTL using the four-liter kettle setup. The last twenty-four melter feed preparation runs were performed at the ACTL using the 22-liter kettle setup. The SRAT rigs were assembled following the guidelines of SRNL-PSE-2006-00074[13]. The intent of the equipment is to functionally replicate the DWPF processing vessels. Each glass kettle is used to replicate both the SRAT and SME, and it is connected to the SRAT Condenser, the Mercury Water Wash Tank (MWWT), and the Formic Acid Vent Condenser (FAVC). The Slurry Mix Evaporator Condensate Tank (SMECT) is represented by a sampling bottle that is used to remove condensate through the MWWT. For the purposes of this paper, the condensers and wash tank are referred to as the offgas components. A sketch of the experimental setup is given as Figure 2-1.

[12] VSL Recipe for Sugar Flowsheet Melter Slurry Prep,

[13] Stone, M. E., Lab-Scale CPC Equipment Set-up, SRNL-ITS-2006-00074, Savannah River Site, Aiken, SC 29808 (2008).

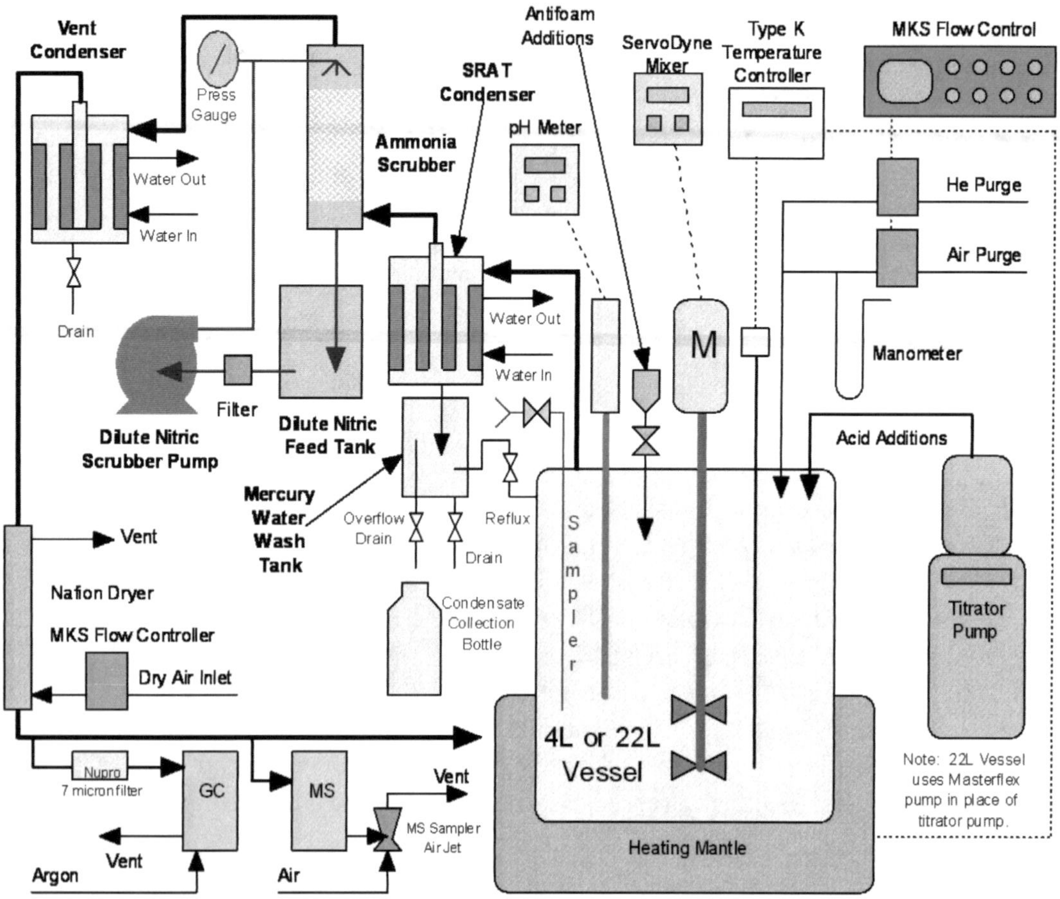

Figure **0-1**. Schematic of SRAT Equipment Set-Up

The runs were performed using the guidance of Procedure ITS-0094 [14] ("Laboratory Scale Chemical Process Cell Simulations") of Manual L29. In runs with noble metals and mercury, offgas hydrogen, oxygen, nitrogen, nitrous oxide, and carbon dioxide concentrations were measured during the experiments using in-line instrumentation. Nitrous dioxide was not measured but was calculated based on oxygen depletion. Helium was introduced at a concentration of 0.5% of the total air purge as an inert tracer gas so that total amounts of generated gas and peak generation rates could be calculated. This approach eliminates the impact of fugitive gas losses through small leaks on the calculated outlet gas flowrates. During the runs, the kettle was monitored to observe reactions that were occurring to include foaming, air entrainment, rheology changes, loss of heat transfer capabilities, and offgas carryover. Observations were recorded on data sheets and pasted into laboratory notebooks.[15, 16]

---

[14] Manual L29, Procedure ITS-0094, Rev. 3, Laboratory Scale Chemical Process Cell Simulations, Savannah River Site, Aiken, SC 29808 (2006)

[15] Glycolic-Formic Flowsheet Development, Part I, WSRC-NB-2011-00021.

[16] Glycolic-Formic Flowsheet Development, Part II, WSRC-NB-2011-00022.

Quality control measures were in place to qualify the data in this report. Helium and air purges were controlled using mass flow controllers calibrated by the SRNL calibration shop using NIST traceable standards and methods. Thermocouples were calibrated using a dry block calibrator. The GCs were calibrated with standard calibration gases. The pH probes were calibrated with pH 4 and pH 10 buffers and rechecked at the conclusion of each run using pH 4, 7 and 10 buffer solutions.
The automated data acquisition system developed for the 4-L SRAT rigs was used to collect data electronically. Data included SRAT temperature, bath temperatures for the cooling water to the SRAT condenser and Formic Acid Vent Condenser (FAVC), slurry pH, SRAT mixer speed and torque, and air and helium purge flows (He is used as an internal standard and is set to 0.5% of the nominal SRAT air purge flow). Cumulative acid addition volume data were collected from the automated dispensers using an algorithm that matches the indicated total on the dispenser. Raw GC data were acquired on separate computers dedicated to each instrument.

Dual column Agilent 3000A micro GC's were used on both runs. The GC's were baked out before and between runs. Column-A can collect data related to He, H_2, O_2, N_2, NO, and CO, while column-B can collect data related to CO_2, N_2O, and water. Calibrations were performed using a standard calibration gas containing 0.499 vol% He, 1.000 vol% H2, 20.00 vol% O_2, 51.0 vol% N_2, 25.0 vol% CO_2 and 2.50 vol% N_2O. Instrument calibration was verified prior to starting the SRAT cycle. Room air was used to give a two point calibration for N_2. Calibration status was rechecked following the SRAT cycle.

Concentrated nitric acid (50-wt%), formic acid (90-wt%) and glycolic acid (70 wt%)-formic acid blends were used to acidify the sludge and perform neutralization and reduction reactions during processing. The amounts of acid to add for each run were determined using the Koopman acid equation.[17] The split of the acid was determined using a modified version of the REDOX equation currently being used in DWPF processing.[18] The first principles electron equivalents expression had a new term added for glycolate ion (see below). The REDOX target ($Fe^{2+}/\Sigma Fe$) was 0.2 for the majority of the runs. To account for the reactions and anion destructions that occur during processing, assumptions about nitrite destruction, nitrite to nitrate conversion, formate destruction, and glycolate destruction were made for each run.

To prevent foaming during SRAT processing, 200 ppm IIT 747 antifoam was added before acid addition, 100 ppm was added after nitric acid addition was complete and 500 ppm was added at the completion of formic acid addition. SRAT processing included 12-hours at boiling (dewater time plus reflux time). The SME processing did not include the addition of canister dewaters. The frit addition was split into two equal portions. The frit was added with water and formic acid at DWPF prototypical conditions. Concentration was performed after each frit addition and then heat was removed to allow for the next frit addition. A final concentration was performed at the end of the run to meet the 50 weight percent total solids target. The SRAT condenser was maintained at 25° C during the run, while the vent condenser was maintained at 4° C.

---

[17] Koopman, D.C., A.I. Fernandez, B.R. Pickenheim, Preliminary Evaluations of Two Proposed Stoichiometric Acid Equations, Revision 0, Savannah River Site, Aiken, SC 29808 (2009).

[18] Jantzen, C.M., J.R. Zamecnik, D.C. Koopman, C.C. Herman, and J.B. Pickett, Electron Equivalents Model for Controlling Reduction-Oxidation (REDOX) Equilibrium during High Level Waste (HLW) Vitrification, WSRC-TR-2003-00126, Savannah River Site, Aiken, SC 29808 (2003).

In runs with noble metals and mercury, a standard 4-L SRAT/SME apparatus with an ammonia scrubber was used for these simulations. The scrubber solution consisted of 749 g of de-ionized water and 1 g of 50 wt% nitric acid. The solution was recirculated through the column by a MasterFlex pump at 300 mL/min through a spray nozzle at the top of the packed section. Glass rings were used as packing and did not significantly add to the back pressure on the SRAT vessel as has been seen in earlier tests with different packing.

Standard SRAT acid calculations were performed with a few modifications. The Koopman minimum acid equation was used with a 125% stoichiometric factor for the first set of all tests. The acid mix was partitioned between nitric and the formic/glycolic blend by utilizing the latest REDOX equation[19] with a term added for glycolate. A coefficient of 6 was used on the glycolate term based on electron equivalence. The REDOX target for these runs was 0.2.

$$REDOX=0.2358+0.1999*((2*C_{formate}+4*C_{oxalate}+4*C_{Carbon}+6*C_{glycolate}-5*(C_{Nitrate}+C_{Nitrite})-5*C_{Mn}))*(45/TS)$$

Where C = species concentration, g-mole/kg melter feed, TS = total solids in melter feed in wt %, and REDOX is a molar ratio of $Fe^{2+}/\Sigma Fe$

Process assumptions were made to predict SME product anion concentrations. In addition to the standard assumptions needed for formate loss and nitrite to nitrate conversion, a factor was added to the acid calculation for glycolate loss. Process assumptions for the stoichiometric window testing were adjusted based on results from the first set of simulations.

CPC Run Details

The first twenty-two glycolic-formic flowsheet tests were performed at the ACTL using the four-liter kettle setup. The first four simulations (GF1-4) were developed to compare the existing nitric/formic flowsheet to the glycolic/formic flowsheet. The second set of four runs (GF5-8) was designed to determine the processing window of the glycolic-formic flowsheet by adjusting acid stoichiometry from 100% to 200% and testing the flowsheet without formic acid (GF8). The third set of four runs (GF9-12) was designed to determine how the form of the catalyst Ru impacted hydrogen generation. These runs are not pertinent to the flowsheet study so will not be discussed further in this report. The fourth set of four runs (GF13-16) was designed to demonstrate that glass produced from the formic-glycolic flowsheet did not have a steep, titration like change that would make REDOX control difficult. The fifth set of runs (GF17-20) was designed to determine what impact the glycolic/formic ratio would have on processing and especially hydrogen generation. The last two runs were designed to develop the processing parameters for melter feed preparation for VSL melt rate testing, before producing large batches of melter feed for the nitric/formic flowsheet (GF21) and the glycolic/formic flowsheet (GF22). The last twenty four melter feed preparation runs, three sets of eight runs, were performed at the ACTL using the 22-liter kettle setup to produce approximately 100 kg of SRAT product for VSL testing of the nitric/formic flowsheet (GF23A-H), the glycolic/formic flowsheet

[19] Jantzen, C.M. and M.E. Stone, Role of Manganese Reduction/Oxidation (REDOX) on Foaming and Melt Rate in High Level Waste Melters, WSRC-STI-2006-00066, Savannah River Site, Aiken, SC, March 2007.

(GF24A-H) and the sugar flowsheet (GF25A-H). Table 2-1 identifies each run and its corresponding assumptions.

Table **0-1**. CPC Simulation Process Assumptions

Run	Objective	Cycles	Date	Acid Stoichiometry	Glycolic % moles	Formic % moles
GF1	Formic Baseline	SRAT/SME	17-May-10	125%	0	100
GF2	Glycolic Baseline No Hg	SRAT/SME	17-May-10	125%	80	20
GF3	Glycolic Baseline	SRAT/SME	19-May-10	125%	80	20
GF4	ARP	SRAT/SME	19-May-10	125%	80	20
GF5	150% Acid Stoichiometry	SRAT/SME	7-Jun-10	150%	80	20
GF6	100% Acid Stoichiometry	SRAT/SME	7-Jun-10	100%	80	20
GF7	200% Acid Stoichiometry	SRAT/SME	9-Jun-10	200%	80	20
GF8	125% Acid Stoichiometry	SRAT/SME	9-Jun-10	125%	100	0
GF9	Old Ru	SRAT	14-Jul-10	125%	0	100
GF10	New Ru	SRAT	14-Jul-10	125%	0	100
GF11	Old Ru	SRAT	21-Jul-10	198%	0	100
GF12	New Ru	SRAT	21-Jul-10	198%	0	100
GF13	REDOX Target 0	SRAT	25-Aug-10	100%	80	20
GF14	REDOX Target 0.2	SRAT	25-Aug-10	100%	80	20
GF15	REDOX Target 0.1	SRAT	16-Sep-10	100%	80	20
GF16	REDOX Target 0.3	SRAT	16-Sep-10	100%	80	20
GF17	40:60 Glycolic/Formic	SRAT	28-Sep-10	125%	40	60
GF18	50:50 Glycolic/Formic	SRAT	28-Sep-10	125%	50	50
GF19	70:30 Glycolic/Formic	SRAT	30-Sep-10	125%	70	30
GF20	60:40 Glycolic/Formic	SRAT	30-Sep-10	125%	60	40
GF21	Formic Baseline	SRAT	2-Nov-10	110%	0	100
GF22	80:20 Glycolic/Formic	SRAT	2-Nov-10	110%	80	20
GF23	Baseline Feed for VSL	SRAT	30 Nov to 6 Dec-10	110%	0	100
GF24	GF Feed for VSL	SRAT	8-14 Dec-10	110%	80	20
GF25	Sugar Feed for VSL	SRAT	8-14 Dec-10	73%	0	100

DWPF design basis processing conditions were scaled down and used for most processing parameters including: SRAT/SME air purges, acid addition rates, and boil-up rate. SRAT product total dried solids were targeted at 25 wt% for the baseline run. Final SME total dried solids were targeted at 45% at 36% waste loading.

The SRAT product solids targets were adjusted for the glycolic/formic flowsheet runs because of the mass of the glycolate ion. Because its molar mass is about 2/3 greater than formic, adding glycolic acid contributes an appreciable amount to SRAT product soluble (and thus total) solids. The glycolic/formic flowsheet SRAT product total dried solids targets were adjusted to higher total solids (same insoluble solids mass) in the SRAT product slurries. No adjustments were made to the SME cycle solids targets.

The melter feed preparation runs (GF21, 22, 23, 24 and 25) used abbreviated processing to allow completion of processing within a twelve-hour shift. To maximize throughput, the nitric and glycolic-formic were fed at the same molar flowrate as formic (a 2x increase in volumetric flowrate for these more dilute acids). In addition, the boil-up rate was set at 100% power to maximize the boil-up rate. Since nitrite was destroyed prior to the concentration endpoint being reached and mercury stripping wasn't needed, the processing was complete when the concentration target was reached (calculated to be 45 wt% solids after the frit was added by VSL). No issues were noted that were associated with the higher acid feed rates. It is recommended that DWPF eventually modify the acid feed pumps to deliver the higher flow rates if the glycolic-formic flowsheet is implemented.

A flowsheet that could meet all the processing constraints would have to be a very robust flowsheet. The following constraints must be met by the DWPF CPC flowsheet:

- SRAT hydrogen <0.65 lb/hr
- SME hydrogen <0.223 lb/hr
- Reduce mercury to elemental form
- Steam strip mercury below 0.6 wt% in the SRAT product dried solids
- SRAT product less than 1000 mg nitrite/kg product slurry
- SRAT product rheology design basis 1.5 to 5 Pa yield stress and 5 to 12 cP consistency
- SME product rheology 2.5 to 15 Pa yield stress and 10 to 40 cP consistency
- Glass REDOX of 0.09-0.33 $Fe^{2+}/\Sigma Fe$
- Minimize water in SME product (45 wt% typical)
- Minimal foaming

Data are presented in Section 3 showing how the glycolic-formic flowsheet met or exceeded the processing constraints in the list above with the possible exception of REDOX.

Process Data Collection

In the 4-L experiments, an automated data acquisition system was used to collect run data every minute on a computer. In all experiments, the process data was manually recorded approximately every twenty minutes including SRAT slurry temperature, slurry pH, cooling water temperatures for the SRAT condenser and Formic Acid Vent Condenser (FAVC), SRAT mixer speed, air and helium purge rates, and raw gas chromatographs.

Agilent 3000 Series GC's were used on all simulations to measure the offgas composition. The GCs were baked out and calibrated with standard calibration gas between runs. Calibration was verified following the completion of the SME cycles.

Analytical Methods

Process samples were analyzed by various methods. Slurry and supernate elemental compositions were measured by inductively coupled plasma-atomic emission spectroscopy (ICP-AES) at the Process Science Analytical Laboratory (PSAL). Soluble anion concentrations were measured by Ion Chromatography (IC). Mercury concentration was measured by ICP-AES. Ammonium ion concentration on selected samples was measured by cation chromatography by SRNL Analytical Development (AD). Slurry and supernate densities were measured using an Anton-Parr instrument at PSAL. Reduction-Oxidation (REDOX) of glasses made from SME product slurries was measured by PSAL. Dewater and condensate samples were submitted to AD for volatile/semi-volatile organics analysis (VOA/SVOA).

A successful gradient method using the AG-11HC and AS-11HC, 2mm microbore columns was developed to run samples for the Alternative Reductant Demonstrations. The method provides good peak resolution for all nine anions. The method that will be used to analyze fluoride, glycolate, formate, chloride, nitrite, nitrate, sulfate, oxalate and phosphate on SRAT/SME samples is as follows (Table 2-2):

*Table **0-2**. Final Ion Chromatography Method for Glycolate and Other Anions*

Instrument	DX500
Columns	2mm AG-11HC, 2mm AS-11HC
Suppressor	ASRS-300, 2mm
Carbonate Removal Device	CRD-20, 2mm
Injection Loop	25 microliter
Calibration Standards	Three point calibration (0.1, 0.5, 1 ppm fluoride, 1, 5, 10 ppm for all other anions)
Gradient Run, Concentration (NaOH)	Time
1.8mM → 30mM Ramp	0-40 minutes
30mM → 1.8mM Ramp	40-50 minutes
Detection Limit	Lowest calibration standard

<u>Issues with analytical method</u>

Organic anions at low concentrations (1 ppm, 5 ppm, 10 ppm) degrade over a short period of time. Due to the potential degradation, it is our recommendation to keep all organic anion standards (manufacturer and calibration) refrigerated and in the dark. New calibration standards should be made every time the instrument is calibrated.

The run time is significantly long due to the request for phosphate. The stability of phosphate when analyzing SRAT and SME samples is extremely sensitive due to the high transition metals (i.e. iron) that build up in the columns. Phosphate has a tendency to bind with the transition metals creating erroneous readings. Also, the transition metals can create peak tailing. This often creates poor accuracy of calibration check standards at the end of sample runs. Phosphate is never above detection limit of the IC for typical SRAT/SME samples and is reported as a less than value. A recommendation is to obtain phosphorus by ICP-AES and remove phosphate from the list of requested anions. This could decrease run time significantly.

Another solution to significantly decrease run time is for DWPF to purchase a new ICS-5000 system that could run multiple methods simultaneously on one sample injection. For example, one method could be set up to elute the first three anions (fluoride, formate and glycolate) slowly off the column to obtain results and another method run to quickly elute these three anions off the column (no peak separation obtained) and then get good separation for the other six anions in 10- 20 minutes. The same solution could be obtained on DWPF's two ICS-3000 systems.

The potential of fluoride interfering with glycolate can be a problem. Over time, baseline peak separation can diminish and a large fluoride peak in the calibration standards can create error in the glycolate analyses, if fluoride is not present. Fluoride in SRAT/SME samples is always below detection. A recommendation is to eliminate this peak to decrease the potential analytical error with glycolate.

Results and Discussion

Primary SRAT/SME simulation analytical data from the glycolic-formic acid flowsheet testing will be presented in the following sections and supplemental data will be included in the Appendices as necessary. The testing is divided into the following segments:

1. Feasibility of Flowsheet (GF1 to GF8)
   a. Flowsheet development GF1 to GF4 and GF8
   b. Acid Stoichiometry Window (GF3, GF5 to GF7)
2. Ruthenium Catalyst Testing (GF9 to GF12). Data will not be presented as it is not pertinent to decision.
3. REDOX Target Testing (GF13 to GF16)
4. Optimum Glycolic-Formic Ratio (GF1, GF17, GF18, GF20, GF19, GF8)
5. Melter Feed Preparation (GF21 to GF25)

The focus of the study was on proving the glycolic/formic flowsheet could meet all current CPC processing objectives, most notably effective removal of mercury in the SRAT while simultaneously reducing hydrogen generation. Much of the data presented here will be a comparison between the two baseline flowsheet cases, GF1 and GF3. These are the only runs performed under the same conditions with both flowsheets and therefore the only direct point of comparison. Data for GF6, the 100% stoichiometry nitric/glycolic/formic run, will also be presented because it appears that running at lower acid stoichiometry is more appropriate for the new flowsheet.

Small-scale Laboratory Results (SRNL)

Twenty-two 4-L Laboratory Scale experiments (GF1 to GF22) were completed to develop the optimum glycolic-formic acid flowsheet for DWPF. Twenty-four 22-L Pilot Scale experiments (GF23A-H, GF24A-H and GF25A-H) were performed to prepare melter feed for VSL testing.

The testing was completed under very challenging conditions. First, the testing used a SB6 slurry produced for SRNL by Harrell Industries. This slurry was very rheologically viscous. During SRAT processing, the pH dropped from about 13 to pH 3 or 4. The slurry actually thickened as the pH dropped to near neutral (pH 7) so mixing of this slurry was very challenging. In addition, conservative concentrations of mercury and noble metals were added. This led to high hydrogen and ammonia generation along with the depletion of anion species (nitrate and formate) that are important to glass REDOX.

SRAT Offgas

Besides essentially eliminating hydrogen generation, the glycolic/formic acid flowsheet also appears to stop or significantly slow down other off-gas generating reactions. The graphs below compare N_2O (Figure 3-1) and CO_2 (Figure 3-2) generation rates for the two baseline flowsheet cases (GF1 for the nitric/formic flowsheet and GF3 for the glycolic-formic flowsheet. The rapid fluctuations, especially in the nitric/formic case, can be attributed to the mixing issues experienced during acid addition.

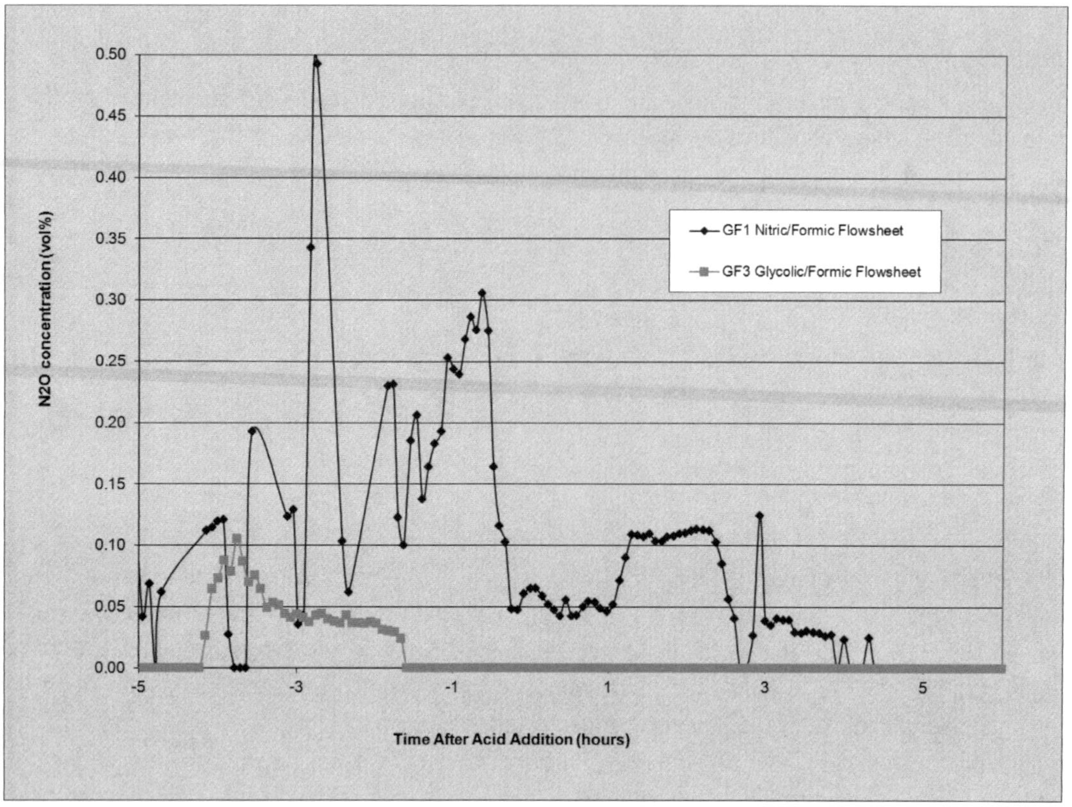

Figure **0-1**. SRAT cycle N_2O generation

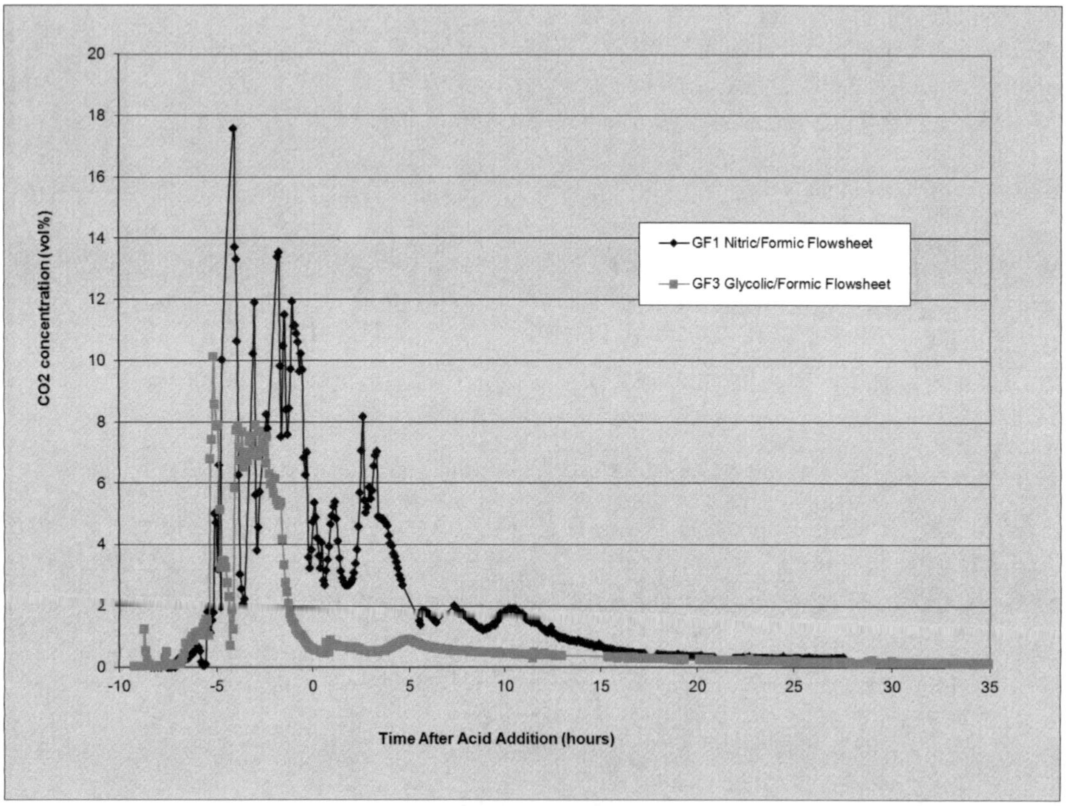

Figure 0-2. SRAT cycle CO_2 generation

It is worth noting that some of the GC data were impacted by mixing issues (meaning peaks on the DWPF flow versus time graphs were less intense than otherwise). Additional off-gas data for other GF simulations can be found in Appendix A3 or in the preliminary reports. Lower generation of N_2O and H_2 were noted in the nitric/glycolic/formic flowsheet runs. GF6 (Figure 3-3) is shown below as an example.

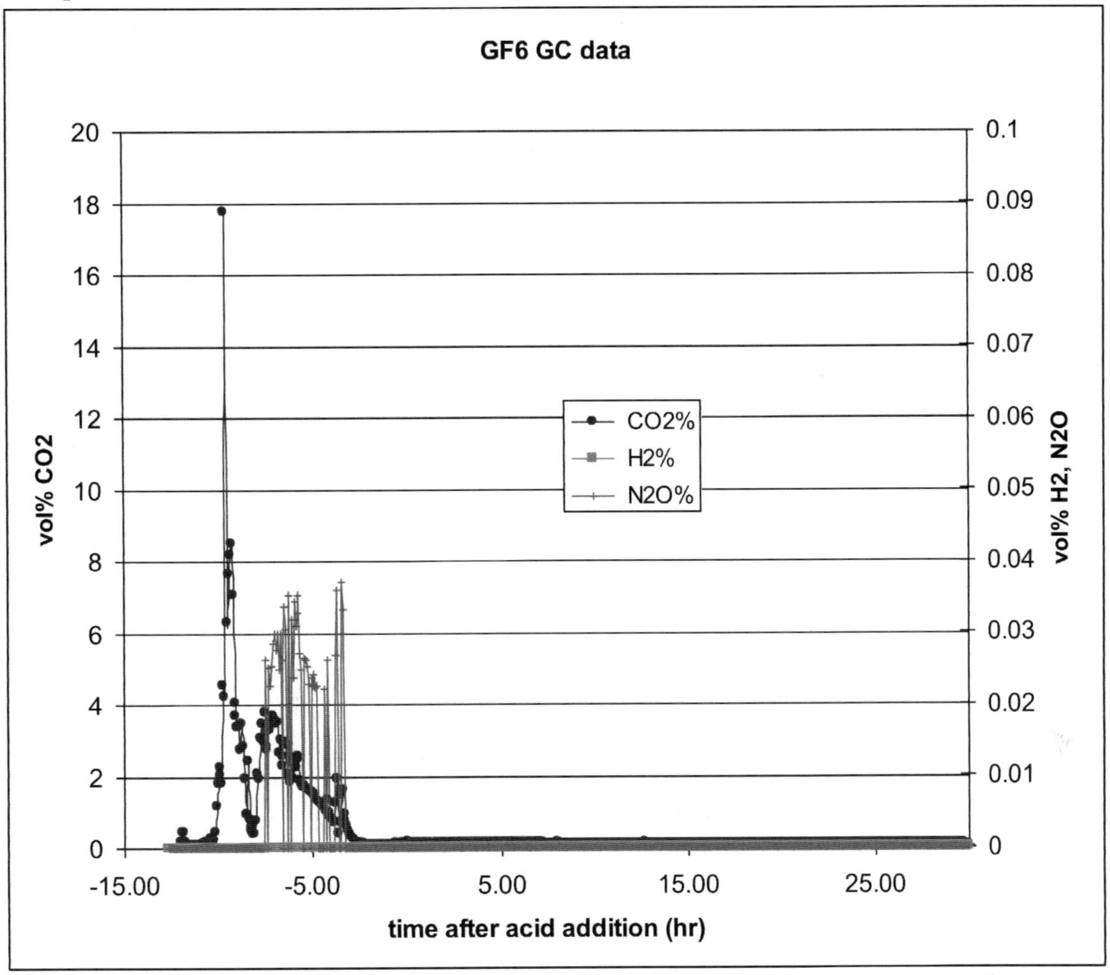

Figure **0-3**. GF6 GC Data

SRAT Chemistry

The chemistry in the glycolic-formic SRAT cycle was very similar to the chemistry in the baseline flowsheet. However, there are a few areas where the chemistry is significantly different and these will be described below:

- The hydrogen generation in the glycolic-formic flowsheet was significantly lower compared to the baseline flowsheet. This will be discussed in Section 3.1.3.
- The formate destruction in the glycolic-formic flowsheet was significantly lower compared to the baseline flowsheet. This will be discussed in Section 3.1.2. This led to much lower carbon dioxide generation in the glycolic-formic flowsheet.
- The glycolic-formic flowsheet is much better at dissolving metals compared to the baseline flowsheet. This led to substantially lower yield stress SRAT and SME products but also led to

significant dissolution of metals such as iron and gadolinium which have been used to provide criticality control. This will be discussed in Section 3.1.2.3.

- There are several flowsheet controls that could be useful in processing future sludge batches. First, a ratio of 4 moles of glycolic per mole of formic was used throughout much of the testing. However, this ratio could be adjusted as needed to optimize processing. For sludge batches with higher mercury or noble metals, using glycolic acid without formic acid would minimize the hydrogen generation while still reducing the mercury content of the SRAT product to meet target. In rheologically challenging slurries, higher acid stoichiometry and a higher glycolic ratio would produce a rheologically thinner product. However, since lower acid stoichiometries exhibited acceptable rheological products, targeting a lower acid stoichiometry (possibly lower than 100% Koopman Stoichiometry) likely will produce optimum processing conditions.

SRAT Product Data

General SRAT product data for the first eight runs are tabulated below (Table 3-1).

Table **0-1**. SRAT Product Anions, Solids, **Density**, and pH Data

anions (mg/kg)	GF1	GF2	GF3	GF4	GF5	GF6	GF7	GF8
fluoride	<100	<100	<100	<100	<100	<100	<100	<100
chloride	1085	1040	976	839	1002	1230	890	1210
nitrite	<100	<100	<100	<100	<100	<100	<100	<100
nitrate	25000	51600	55950	53100	59900	42250	59900	56250
sulfate	1920	1680	1765	2120	3275	2565	3280	3485
glycolate	<100	53950	55400	56850	114500	65050	161500	118500
oxalate	296	649	2405	3910	2775	3395	1645	4680
formate	12900	2555	<100	<100	824	776	3750	<100
phosphate	<100	<100	<100	<100	<100	<100	<100	<100
wt% total solids	17.53%	21.77%	21.79%	22.32%	24.03%	25.59%	21.75%	24.62%
wt% calcined solids	12.19%	11.37%	11.36%	11.86%	12.02%	14.13%	10.20%	12.47%
wt% insoluble solids	11.01%	8.76%	8.98%	8.84%	10.23%	11.20%	5.96%	10.83%
wt% soluble solids	6.53%	13.01%	12.80%	13.48%	13.80%	14.39%	15.79%	13.80%
density (g/mL)	1.084	1.163	1.158	1.161	1.166	1.187	1.138	1.178
pH at 25°C	4.24	4.01	3.99	4.07	3.17	5.04	3.01	3.02

The oxalate results are of particular interest. The starting sludge contains about 800 mg/kg oxalate, which could be partially destroyed catalytically during the SRAT cycle. In the glycolic/formic flowsheet runs, however, oxalate is being created. It is possible that glycolic acid is oxidized to glyoxylic acid ($HCOCO_2H$) by nitrite or MnO_2, which is further oxidized to oxalic acid by the reduction of mercury. This also would explain the lack of oxalate generation in the run without mercury, GF2.

These runs represent the first time that glycolate ion measurements were performed on actual SRAT/SME products, as opposed to spiked samples and simple solutions. A separate report has been drafted detailing the glycolate IC method development.[20] It appears from the data presented here that the second set of runs (GF5-8) may be biased high in glycolate. The 65,050 mg/kg measured for GF6 represents 99.6% of the total moles of glycolate added during the SRAT. This is unrealistically high, especially if the proposed pathway for oxalate formation is correct. The results for GF5-8 show more glycolate in the SRAT product than was added. Resolving this issue so that reliable glycolate data is available will be especially important for refining the REDOX model as flowsheet development continues.

Anion balance data for nitrite, nitrate, formate and glycolate are presented in the table below for runs GF1-4 (Table 3-2).

Table **0-2**. Anion Balance Data

	GF1	GF2	GF3	GF4	GF5	GF6	GF7	GF8
SRAT Formate Destruction (%)	78.7	71.4	100	100	94.0	92.1	71.3	N/A
SRAT Nitrite Destruction (%)	100	100	100	100	100	100	100	100
SRAT Glycolate Destruction (%)	N/A	9.3	11.7	4.5	-24	0.4	-85	-80
SRAT Nitrite to Nitrate Conversion (%)	18.9	44.4	51.4	46.4	-0.2	27.1	-3.2	73.5
SME Formate Destruction (%)	-3.5	44.3	100	41.5	61.8	42.6	62.2	N/A
SME Nitrate Destruction (%)	25.6	9.7	9.3	9.9	-2.5	10.1	21.8	N/A
SME Glycolate Destruction (%)	N/A	5.4	9.3	5.3	15.1	4.9	35.8	N/A

SRAT (Table 3-3) and SME (Table 3-4) product elemental data are summarized in the following tables.

[20] Best, D.R., Anion Analysis by Ion Chromatography for the Alternate Reductant Program for the Defense Waste Processing Facility, SRNL-STI-2010-00389, Savannah River National Laboratory, Aiken, SC, June 2010.

Table **0-3**. **SRAT Product Elemental Data**, wt % calcined solids basis

element	GF1	GF2	GF3	GF4	GF5	GF6	GF7	GF8
Al	16.5	16.3	16.1	14.2	16.4	16.3	16.7	16.6
B	<0.100	0.1	<0.100	<0.100	<0.100	<0.100	<0.100	<0.100
Ba	0.07	0.04	0.05	0.08	0.04	0.04	0.04	0.04
Ca	0.60	0.53	0.54	0.66	0.57	0.59	0.57	0.56
Cd	0.08	0.04	0.04	0.03	<0.010	<0.010	<0.010	<0.010
Ce	<0.010	<0.010	<0.010	<0.010	<0.010	<0.010	<0.010	<0.010
Cr	0.12	0.12	0.12	0.11	0.13	0.13	0.13	0.16
Cu	0.23	0.25	0.23	0.19	0.25	0.16	0.27	0.26
Fe	19.2	19.3	19.2	17.1	18.6	18.4	18.7	18.7
K	0.04	0.17	0.03	0.03	0.03	0.03	0.06	0.03
La	<0.010	<0.010	<0.010	<0.010	<0.010	<0.010	<0.010	<0.010
Li	<0.100	0.06	<0.100	<0.100	<0.100	<0.100	<0.100	<0.100
Mg	0.46	0.46	0.46	0.50	0.47	0.47	0.48	0.48
Mn	6.44	6.29	6.48	5.39	6.29	6.47	6.34	6.39
Mo	<0.010	<0.010	<0.010	<0.010	<0.010	<0.010	<0.010	<0.010
Na	15.55	15.45	15.60	16.45	15.69	16.44	16.03	15.99
Ni	2.68	2.70	2.69	2.16	2.63	2.65	2.66	2.68
P	<0.010	<0.010	<0.010	<0.010	<0.010	<0.010	<0.010	<0.010
S	0.43	0.46	0.48	0.53	0.46	0.47	0.46	0.42
Si	0.22	0.17	0.16	0.32	0.20	0.16	0.21	0.23
Sr	0.03	0.03	0.03	0.02	0.08	0.08	0.08	0.08
Ti	0.02	0.02	0.02	3.13	0.01	0.01	0.01	0.01
Zn	<0.010	<0.010	<0.010	<0.010	0.01	0.01	0.01	<0.010
Zr	0.25	0.24	0.24	0.23	0.26	0.25	0.26	0.23

Note – GF4 is the run with ARP/MCU and contains MST as a source of titanium

Table 0-4. SME **Product Elemental Data**, wt % calcined solids basis

Element	GF1	GF2	GF3	GF4	GF5	GF6	GF7
Al	6.17	6.02	5.87	5.13	6.03	5.71	6.08
B	1.44	1.52	1.45	1.48	1.26	1.22	1.20
Ba	0.03	0.01	0.02	0.03	<0.010	<0.010	<0.010
Ca	0.15	0.17	0.16	0.23	0.23	0.24	0.24
Cd	0.03	<0.010	<0.010	<0.010	<0.010	<0.010	<0.010
Ce	<0.010	<0.010	<0.010	<0.010	<0.010	<0.010	<0.010
Cr	0.06	0.05	0.05	0.05	0.05	0.05	0.05
Cu	0.11	0.09	0.09	0.09	0.07	0.05	0.15
Fe	7.27	6.77	7.04	6.08	7.35	6.86	6.78
K	0.06	0.12	0.05	0.05	0.06	0.05	0.10
La	<0.010	<0.010	<0.010	<0.010	<0.010	<0.010	<0.010
Li	2.15	2.28	2.19	2.21	2.22	2.15	2.12
Mg	0.19	0.19	0.18	0.20	0.20	0.20	0.20
Mn	2.31	2.26	2.22	1.83	2.58	2.41	2.32
Mo	<0.010	<0.010	<0.010	<0.010	<0.010	<0.010	<0.010
Na	9.71	9.78	9.37	10.15	9.39	9.53	9.69
Ni	0.87	0.75	0.75	0.60	0.80	0.81	0.93
P	0.02	<0.010	<0.010	<0.010	<0.010	<0.010	<0.010
S	0.16	0.14	0.14	0.14	0.16	0.15	0.14
Si	22.92	23.69	23.69	23.67	23.46	23.71	23.02
Sr	<0.010	<0.010	<0.010	<0.010	0.03	0.03	0.04
Ti	0.04	0.05	0.04	1.02	0.03	0.03	0.03
Zn	<0.010	0.01	<0.010	0.01	<0.010	<0.010	<0.010
Zr	0.17	0.17	0.16	0.24	0.20	0.19	0.19

SRAT Dissolution of Metals as a function of acid stoichiometry

Glycolic acid is currently used commercially to descale heat exchanger surfaces (DuPont uses a mixture of 2 parts glycolic acid and 1 part formic acid) and to dissolve rust. So it is not surprising that the glycolic acid dissolves a number of metals in sludge. Glycolic acid is added to paints and emulsions to give them "better flow properties". Dissolution of metals can be a good thing as it can greatly lower the yield stress of the slurry and allow further concentration of the slurry. As a result, the use of glycolic acid has the potential to allow processing of slurries that could not be processed with the baseline flowsheet. In our testing, the glassware stayed much cleaner in the experiments with the glycolic-formic flowsheet. This was especially true in the melter feed preparation runs where four runs were completed sequentially in the same 22-L glass vessel with no cleaning between runs.

However, the higher the glycolic to formic ratio, the more metals were dissolved. The addition of glycolic acid in the glycolic-formic flowsheet runs caused some elements that are traditionally insoluble to become soluble. Al, Ca, Cu, Fe, Mg, Mn, Ni, Si, Sr, and Zr are much more soluble in the formic-glycolic flowsheet runs. Of particular concern is iron and gadolinium solubility because criticality control at DWPF is based on iron and plutonium not partitioning between the solids and supernate. The chart below (Figure 3-4) shows iron solubility as a function of acid stoichiometry for the nitric/glycolic/formic acid flowsheet (GF3) and the baseline nitric/formic run (GF1) as a point of comparison. It should be noted that metals such as Fe and Gd dissolve and reprecipitate during processing so the percent soluble can change throughout a run.

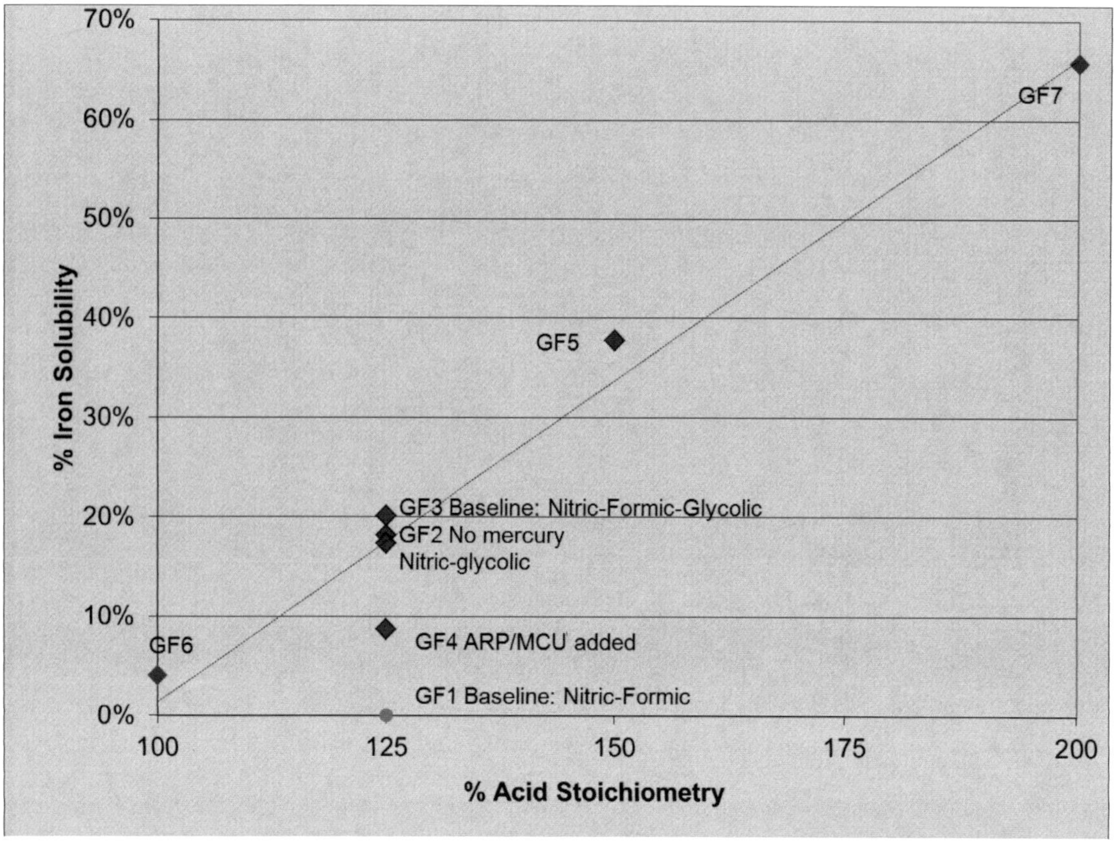

Figure **0-4**. Iron solubility as a function of acid stoichiometry

Other elements also show significantly higher solubility in the glycolic/formic flowsheet runs. The table below shows the amount of each element found in the SRAT product supernate expressed as a percentage of the total element present. These data are calculated by dividing the supernate concentration by the total slurry fraction of each element. Numbers greater than 100% are not physically possible and are a result of error in one of the analytical measurements used in the calculation. Note that the table is grouped to put the acid stoichiometry in order to compare the solubility with stoichiometry. Also, the GF4 data (ARP run) and GF8 data (100% glycolic acid run) are grouped at the bottom of the table (Table 3-5).

Table 0-5. Acid Stoichiometry Study: SRAT Product Supernate % of Element Dissolved

Run	% Acid	Al	Ca	Cu	Fe	Mg	Mn	Ni	Si	Sr	Zr
Baseline	125	0.0 %	8.8 %	0.0 %	0.0 %	1.8 %	0.0 %	0.0 %	22.6 %	0.0 %	0.0 %
GF6	100	2.2 %	116 %	20.8 %	4.1 %	44.0 %	99.0 %	29.3 %	39.1 %	84.7 %	17.7%
GF2	125	4.9 %	131 %	83.5 %	18.1 %	70.1 %	82.6 %	58.9 %	42.1 %	87.8 %	45.7%
GF3	125	4.9 %	128 %	92.0 %	20.1 %	73.2 %	82.1 %	61.7 %	45.2 %	89.9 %	47.7%
GF5	150	5.3 %	130 %	104 %	37.7 %	84.6 %	72.0 %	89.6 %	51.0 %	99.1 %	55.7%
GF7	200	14.5 %	138 %	109 %	65.7 %	92.4 %	91.2 %	92.7 %	117 %	107 %	82.0%
GF4 (ARP)	125	5.9 %	97.6 %	69.2 %	8.7 %	72.9 %	62.4 %	45.8 %	31.1 %	64.1 %	9.8 %
GF8	125	9.4 %	129 %	95.9 %	17.3 %	79.6 %	71.2 %	81.5 %	39.2 %	93.6 %	60.7%

Data for the "baseline" case element dissolutions are not typical of most recent sludge batch testing. This difference is presumably due to the HM levels of noble metals that consumed over 70% of the formic acid in the SRAT and led to a SRAT product pH of about eight. These conditions are typically associated with significant absorption of the CO_2 by-product of formate destruction and subsequent reprecipitation of Ca, Mg, and Mn from the SRAT supernate during reflux.

The zirconium dissolution data in the GF flowsheet runs could impact future SRS accelerated closure projects, such as Small Column Ion Exchange, which use crystalline silicotitanate (CST) ion exchange resin. The binder material in CST contains zirconium, so the GF flowsheet might actively break down the CST binder material in the SRAT.

SRAT Dissolution of Metals as a Function of Glycolic-Formic Acid Ratio

Seven glycolic-formic flowsheet runs were completed to determine the optimum ratio of formic acid to glycolic acid for this flowsheet as this ratio was adjusted from 0% glycolic (baseline) to 100% glycolic (GF8). In earlier testing, glycolic did not reduce mercury. As a result, it was assumed that some formic acid was needed to reduce mercury. Elemental mercury can be steam stripped during SRAT processing so that the SRAT product mercury concentration would be less than the processing limit. However, the mercury target was met, even in the run with no added formic acid (GF8). Therefore any glycolic to formic ratio should be acceptable for mercury removal. Processing at a lower glycolic to formic ratio (more formic acid) leads to higher hydrogen generation. Processing at a lower acid stoichiometry and with a lower glycolic to formic ratio would cause fewer metals to become soluble.

In addition, most of the metals became at least partially soluble as the percentage of glycolic acid increased from 0 to 100%. For example, aluminum went from being essentially insoluble in the baseline flowsheet to dissolving almost 10% of the Al with 100% glycolic acid. Iron went from being essentially insoluble in the baseline flowsheet to dissolving almost 20% of the Fe with 100% glycolic acid. The solubility of many of the elements, including Ca, Cu, Mg, Mn, Ni and Zr increased as the % glycolic acid increased, approaching 100% solubility. As a result, the primary insoluble solids left in the SRAT product were aluminum and iron (Table 3-6, Figure 3-5). This data could be used to optimize the glycolic-formic ratio to minimize hydrogen and iron solubility.

Table 0-6. Glycolic Ratio: SRAT Product Supernate % of Element Dissolved at 125% Koopman Stoichiometry

Run	% Glycolic	Al	Ca	Cu	Fe	Mg	Mn	Ni	Si	Zr
Baseline	0	0.0%	8.8%	0.0%	0.0%	1.8%	0.0%	0.0%	22.6%	0.0%
GF17	40	0.3%	113%	14.2%	0.2%	41.2%	44.1%	0.1%	16.0%	1.4%
GF18	50	0.7%	123%	4%	2.0%	64.8%	84.9%	10.4%	21.4%	3.9%
GF20	60	1.2%	122%	10%	9.4%	61.0%	94.1%	37.2%	25.4%	23.7%
GF19	70	3.0%	123%	62.1%	15.6%	65.8%	92.5%	52.0%	28.6%	42.4%
GF3	80	4.9%	128%	92.0%	20.1%	73.2%	82.1%	61.7%	45.2%	47.7%
GF8	100	9.4%	129%	95.9%	17.3%	79.6%	71.2%	81.5%	39.2%	60.7%

Note that the calcium data in tables 3-5 and 3-6 are quite unusual, but are typical of the results seen in the runs with glycolic acid. Calcium extents of dissolution in the 50-80% range are more typical of sludge batch testing with the formic acid flowsheet. It is not clear at this time whether slurry total calcium may be coming in artificially low or supernate calcium artificially high leading the percent dissolutions in the 120-130% range.

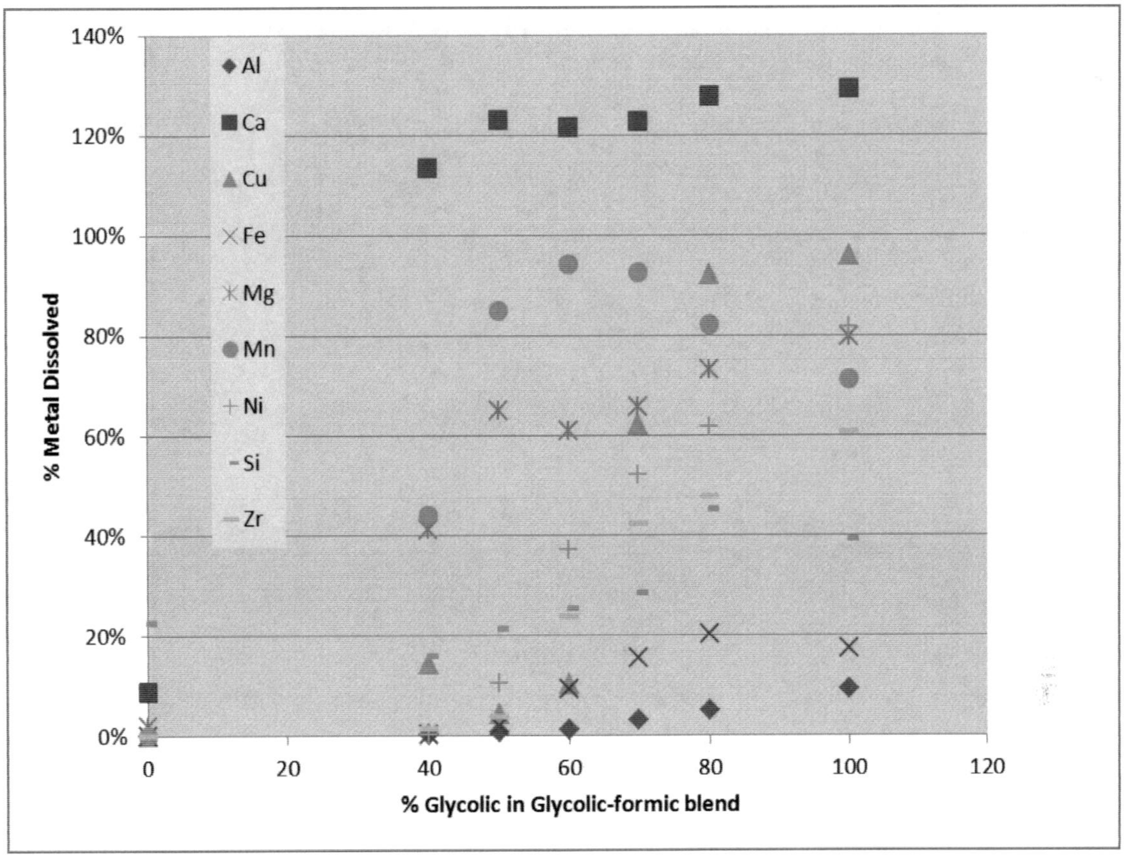

Figure **0-5**. Metal solubility as a function of % glycolic acid in blend

Gadolinium may be used as a substitute neutron poison in the Tank Farm. This would allow the addition of Pu into the sludge batches without causing a potential criticality in DWPF. Testing was completed using the baseline flowsheet and glycolic-formic flowsheet[21] to determine the solubility of metals, including iron and gadolinium.

Gadolinium solubility is related to solution pH (the lower the pH the higher the gadolinium solubility). In the glycolic-formic flowsheet runs with simulant, the SRAT and SME product pH may be less than 4, so >80% of the gadolinium is soluble. Iron in the baseline nitric/formic acid simulant runs is approximately 20% soluble. In the one run with actual waste the iron solubility was close to zero, suggesting that the iron produced in simulant development is more readily solubilized than the iron in the actual waste. Thus, gadolinium may not be an effective neutron poison for the glycolic formic flowsheet. Figure 3-6 and 3-7 summarizes the iron and gadolinium solubility in simulant experiments for both flowsheets. Note that run SB6-29 is a baseline flowsheet run from another study. SB6-29 used the SB6-E sludge simulant, not the SB6-H sludge simulant used for all GF runs[21].

[21] S. H. Reboul, D. R. Best, M. E. Stone, D. R. Click, Partitioning of Gadolinium in the Chemical Processing Cell, SRNL-STI-2010-00804, Savannah River National Laboratory, Aiken, SC, January 2010.

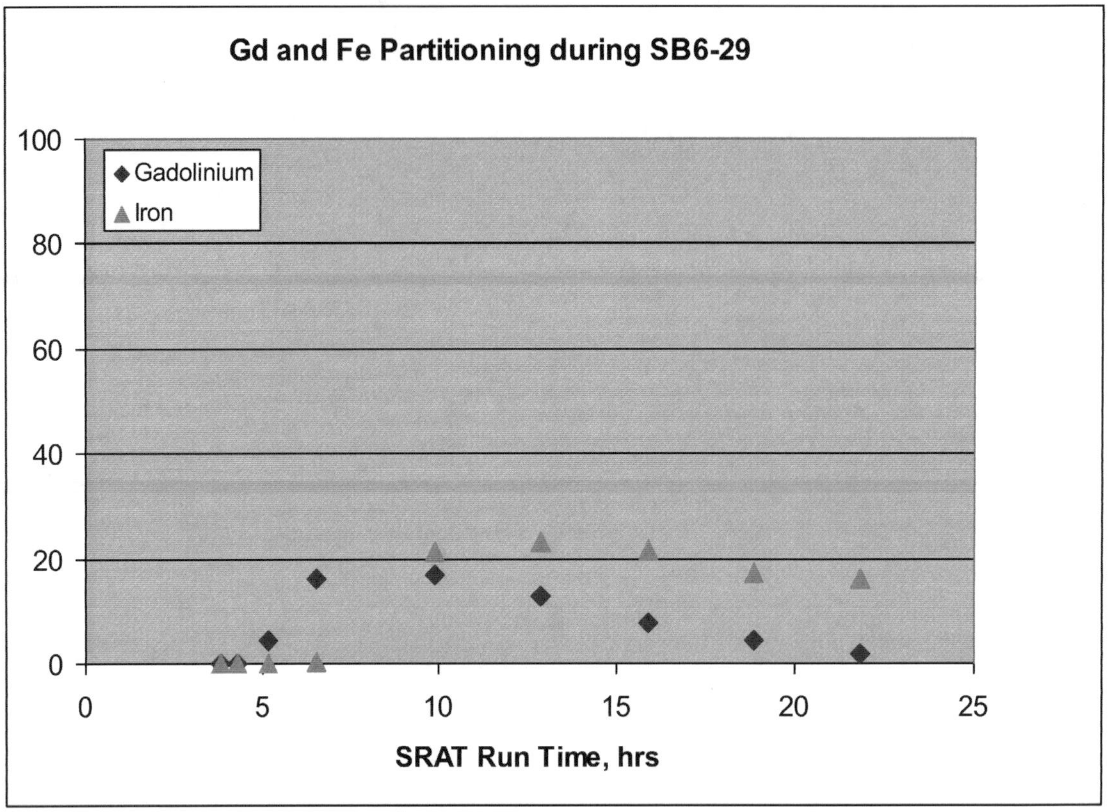

Figure 0-6. Iron and Gadolinium Solubility in SB6-29

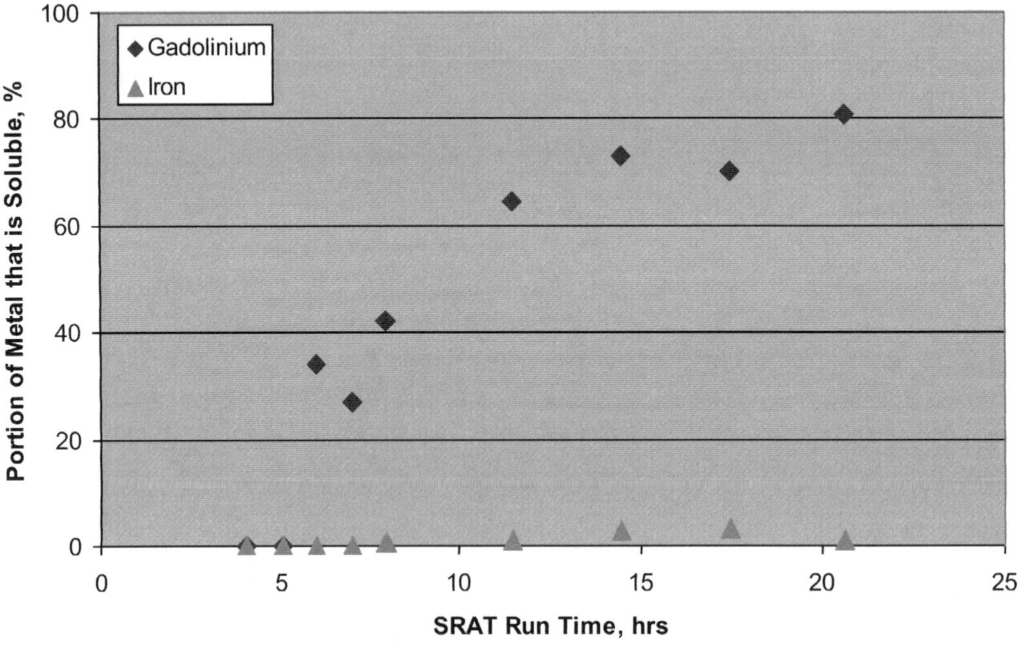

Figure 0-7. Iron and Gadolinium Solubility in GF13

SRAT Degradation Products

Condensate samples were analyzed by SRNL AD for volatile and semi-volatile organics to ensure that there are no new degradation products formed by glycolic acid that could present downstream issues to the Tank Farm. No semi-volatile organics were found in any of the samples. Small amounts (ranging from 5-15 mg/L) of siloxanes, primarily hexamethyldisiloxane, were found in all the samples. This is consistent with antifoam degradation and is not unique to the glycolic/formic acid flowsheet.

Irradiation testing is currently being completed on SRAT simulant supernate to determine whether polymerization of the glycolic acid occurs due to irradiation simulating storage in the DWPF SRAT and SME[22]. Introducing glycolic acid to high radiation doses may cause some polymerization of the glycolic acid in an oxygen depleted atmosphere. Results will be reported when complete.

SRAT Hydrogen Suppression

The main objective of this flowsheet testing was to show that hydrogen generation could be mitigated or eliminated by the use of the glycolic/formic flowsheet. In the first six simulations with the glycolic/formic flowsheet and mercury, hydrogen concentration in the off-gas never exceeded 0.03 volume percent in the SRAT cycle. When formic acid was added with the frit in the SME cycle, hydrogen generation on the order of 0.2 volume percent was noted in the 200% stoichiometry case, GF7. This can be compared to the baseline nitric/formic flowsheet run where hydrogen concentration in the SRAT exceeded the DWPF design basis limit of 0.65 lb/hr (>2 volume percent). Table 3-7 compares SRAT and SME cycle hydrogen on a DWPF scale. In the one run with 100% glycolic acid (no formic acid), no hydrogen was detected at any time throughout the SRAT cycle. This run did not have a SME cycle.

[22] Grant, P.M. and Ward, R.B., Effects of Radiation, Part II. Characterization of the Products from Glycolic Acid, J. Chem. Soc, p 2554, (1959).

Table 0-7. Hydrogen Generation

Run	% Acid Stoichiometry	Note	Glycolic: Formic	SRAT Hydrogen, lb/hr DWPF Scale	SME Hydrogen, lb/hr DWPF Scale
GF1	125	Baseline	0:100	1.614	0.007
GF2	125	No Hg	80:20	0.106	0.171
GF3	125	GF Baseline	80:20	0.006	0.004
GF4	125	ARP/MCU	80:20	0.004	0.003
GF5	150	GF Higher Acid	80:20	0.008	0.013
GF6	100	GF Lowest Acid	80:20	0.013	0.012
GF7	200	GF Highest Acid	80:20	0.024	0.049
GF8	125	100% Glycolic	100:0	0.000	NA
GF13	100	0 REDOX	80:20	0.000	NA
GF14	100	0.2 REDOX	80:20	0.008	NA
GF15	100	0.1 REDOX	80:20	0.005	NA
GF16	100	0.3 REDOX	80:20	0.006	NA
GF17	125	40 % Glycolic	40:60	0.547	NA
GF18	125	50 % Glycolic	50:50	0.198	NA
GF19	125	70 % Glycolic	70:30	0.021	NA
GF20	125	60 % Glycolic	60:40	0.068	NA

The highest hydrogen generation in the Glycolic-formic runs occurred in the runs with the highest formic acid ratio (Runs GF17-20 varied the glycolic-formic ratio). Run GF17 (40% glycolic, 60% formic acid blend) had the highest hydrogen generation at 0.547 lb/hr DWPF scale of any of the glycolic-formic acid runs. The data is summarized in Figure 3-8.

The peak hydrogen generation in the baseline run (GF1) was 260 times higher than in the glycolic-formic run with mercury and noble metals (GF3). The hydrogen profile in the SRAT cycles is summarized in Figure 3-9 for the SRAT cycle and Figure 3-10 for the SME cycle.

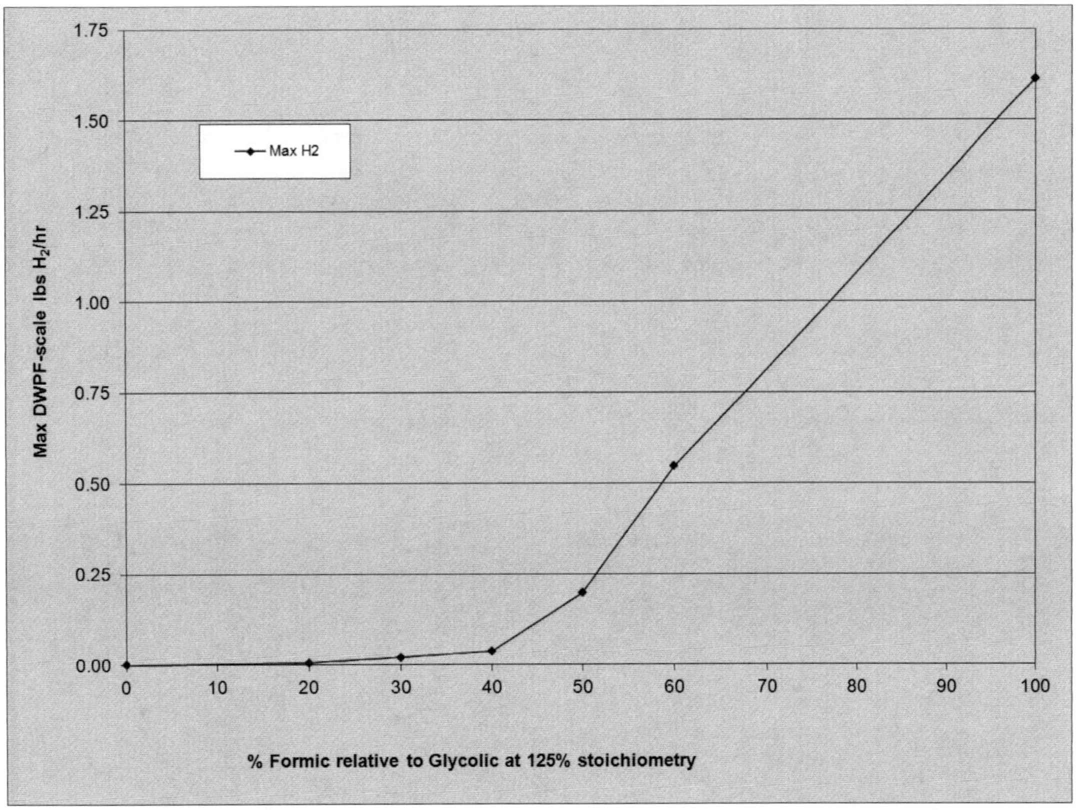

Figure **0-8**. Max Hydrogen Generation Relative to mol% Formic in Formic/Glycolic blend

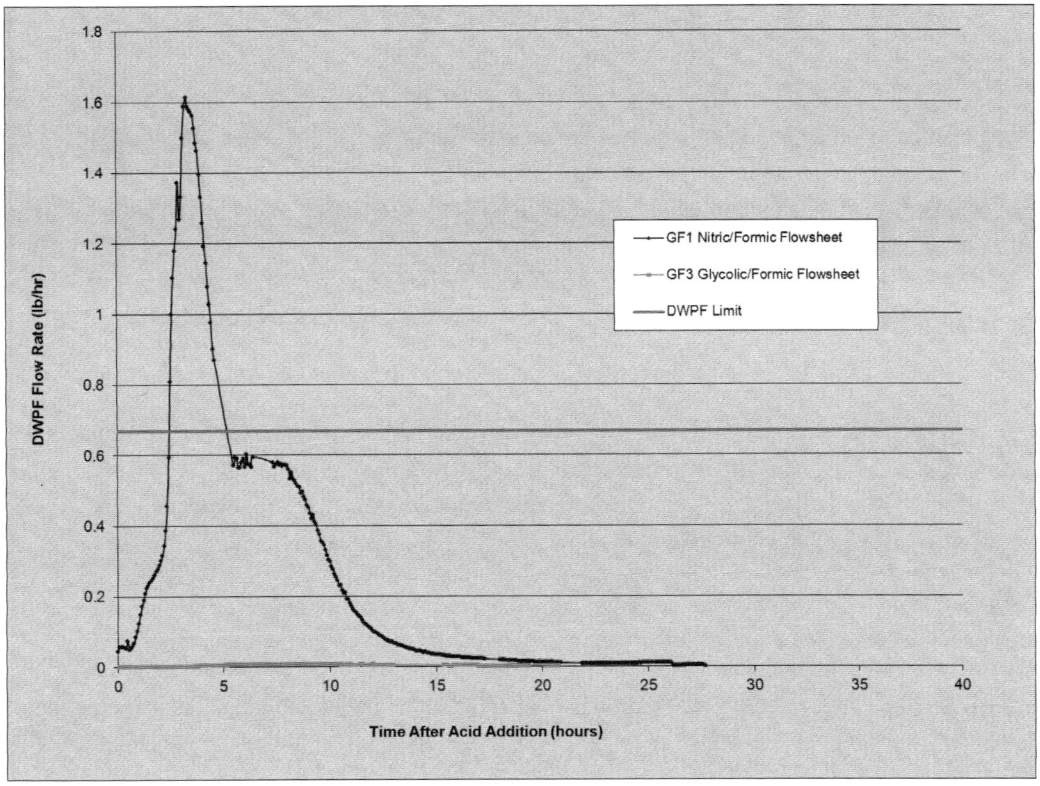

Figure **0-9**. Baseline SRAT cycle hydrogen generation (GF1 vs. GF3)

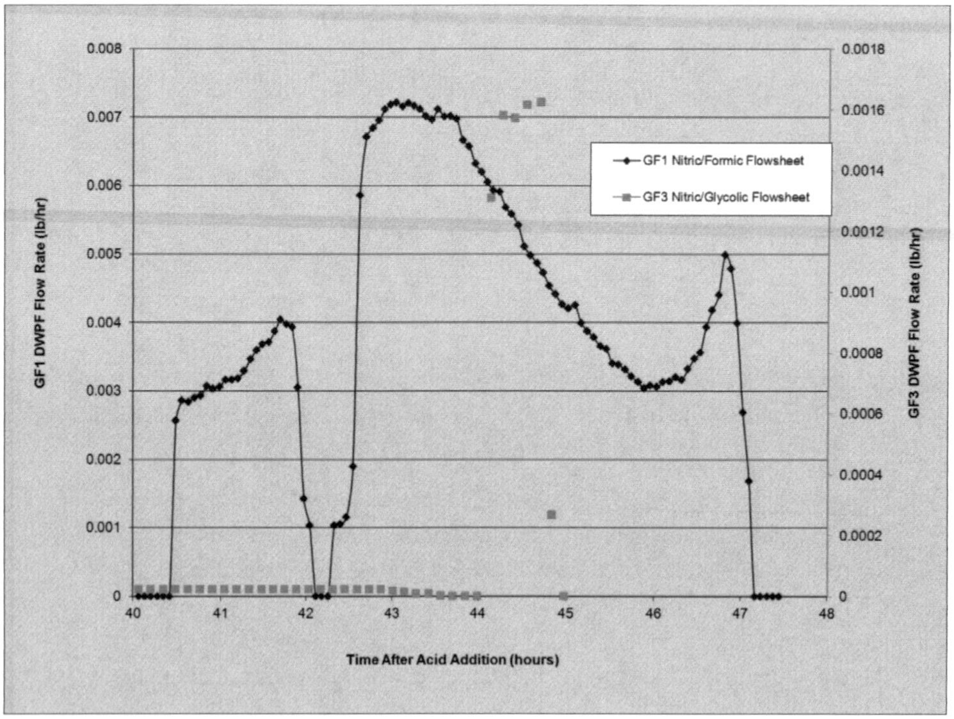

Figure 0-10. Baseline SME Cycle Hydrogen Generation (GF1 vs. GF3)

Without mercury present (GF2), more hydrogen was generated by processing with the glycolic/formic flowsheet (Figure 3-11). This is consistent with testing of the baseline flowsheet[23]. Hydrogen generation was slowly increasing near the end of reflux and appeared to plateau at a maximum of about 0.14 volume percent at the end of the SRAT cycle (peak hydrogen was 17 times higher in run without mercury than in GF3). More hydrogen was generated in the SME cycle; with a maximum near 0.65 volume percent (peak hydrogen was 46 times higher in run without mercury). This is likely in part due to formic acid added with the frit. The hydrogen generation profile for the no mercury run is shown in the figure below.

Note that these runs were performed with HM levels of noble metals and mercury which leads to a high hydrogen generation rate. The maximum hydrogen generation rate is dependent on the noble metal and mercury concentration for a given sludge batch. Actual noble metal concentrations are likely to be significantly lower than were used in this testing.

[23] Koopman, D.C., D.P. Lambert, M.A. Baich, Review Of Catalytic Hydrogen Generation in the DWPF Chemical Processing Cell, Part II, WSRC-TR-2005-00206, Savannah River National Laboratory, Aiken, SC, August 2005.

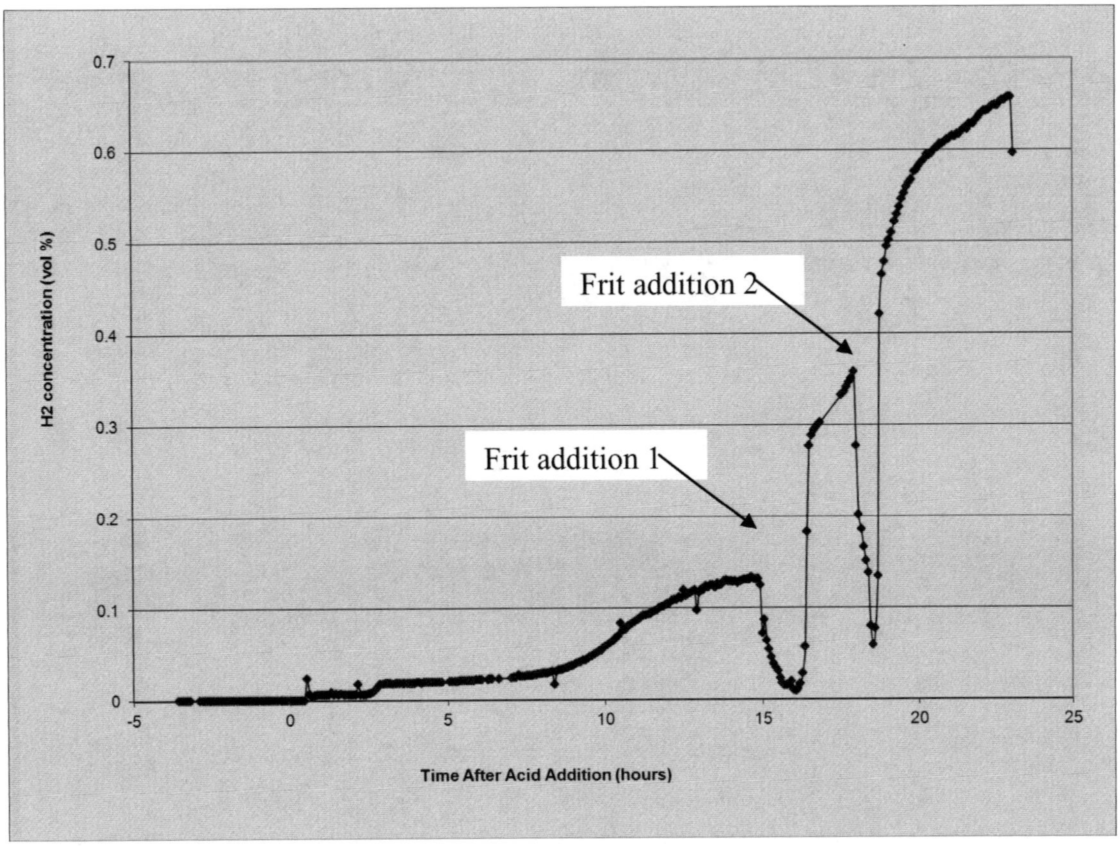

Figure **0-11**. GF2 (no Hg) hydrogen generation

Other graphs of hydrogen, CO_2, and N_2O generation, can be found, i.e. in previously issued reports or in the Appendix A3 of this report.

SRAT Carbon and Nitrogen Balance

A carbon and nitrogen balance was completed. The results for runs GF1 (baseline formic flowsheet) and GF3 (Baseline Glycolic-formic Flowsheet) will be discussed and are included in Appendix A.4. After each run, a SRAT product, SME product and a composite condensate sample were pulled and analyzed for anions and ammonium. The ammonium concentration in the SRAT and SME products was less than the detectable limit (500 or 1000 mg/L). In addition, samples were submitted from the ammonia scrubber liquid (the condensate from the runs was not drained to the SMECT nor recirculated through the scrubber) and FAVC for ammonium analysis. Ammonium ion was found in the condensates and scrubber liquid samples although it was below the detection limit in the slurry samples due to sodium interferences. The inputs to the mass balance are summarized in the Tables 3-8 to 3-10 below.

Table **0-8**. SRAT Carbon and Nitrogen Species in Offgas, g

Run	GF1	GF2	GF3	GF4	GF5	GF6	GF7	GF8
% Glycolic:% Formic	0:100	80:20	80:20	80:20	80:20	80:20	80:20	100:0
Acid Stoichiometry, %	125	125	125	125	150	100	200	125
Nitrous Oxide (N_2O)	1.15	0.50	0.13	0.19	0.14	0.49	0.01	0.00
Nitrogen Oxide (NO_2)	93.4	44.7	84.4	52.7	25.0	35.2	45.5	NA
Nitric Oxide (NO)	0.08	0.99	0.05	1.06	0.07	2.63	1.60	14.6
Carbon Dioxide	105.0	33.4	38.5	35.7	40.7	37.4	62.4	0.076

Table **0-9**. Carbon and Nitrogen Species in Sludge, g

Run	All Runs
TIC	2.82
Oxalate	2.30
Nitrate	17.33
Nitrite	47.32

Table **0-10**. SRAT Carbon and Nitrogen added as Acids, g

Run	GF1	GF2	GF3	GF4	GF5	GF6	GF7	GF8
Formic Acid	182.71	26.74	26.51	24.09	30.64	23.01	41.36	0.00
Nitric Acid	41.38	112.85	112.66	94.00	143.16	77.76	187.45	125.82
Glycolic Acid	0.00	176.71	175.19	159.25	202.54	152.10	273.36	203.11

SRAT Cycle Carbon Balance

A number of carbon species are present in the sludge simulant (TIC, TOC, and oxalate). Carbon is added during SRAT processing as formic acid, glycolic acid, and antifoam. The carbon added with the antifoam will be ignored as it is a minor contributor. Carbon is also added during the SME processing as formic acid although this again is a minor contributor. The organic acids can decompose to smaller organic species such as formate, oxalate, carbonate, CO and CO_2. Instrumentation to quantify CO was not available for this testing. The primary source of carbon in both flowsheets is the formic and glycolic acid added. The overall carbon balance is summarized in Table 3-11.

Table **0-11**. Carbon and Nitrogen Overall Balance

Run	GF1 C	GF3 C	GF1 N	GF3 N
Sludge, g	3.8	3.8	20.1	20.3
Acid, g	47.6	67.4	22.5	25.0
Total In, g	**51.4**	**71.3**	**42.6**	**45.3**
Offgas, g	29.7	11.0	19.1	11.7
Product , g	8.3	46.6	11.5	28.8
Condensate, g	0.2	0.2	0.1	0.1
Total Out, g	**38.2**	**57.8**	**35.1**	**40.8**
% Closure	74	81	82	90

SRAT Cycle Nitrogen Balance

A number of nitrogen species are present in the sludge (nitrite, nitrate). Nitrogen is added during SRAT processing as nitric acid. No nitrogen is added during the SME cycle. The major decomposition species of nitrate and nitrite include nitrous oxide, nitric oxide, nitrogen oxide, and ammonium. In addition, nitrogen gas could be produced but there is no way to quantify this using an air purge as this is a small quantity compared to the nitrogen present. In addition, NO_2 cannot be measured with the GC so its concentration was calculated by assuming that 2 moles of NO_2 were produced for each mole of oxygen consumed. It is obvious in these runs that a large quantity of NO_2 is produced since the offgas turns orange during acid driven nitrite destruction. However, there are other reactions that consume oxygen so this method may overestimate the NO_2 production. The overall carbon balance is summarized in Table 3-11. The baseline formic flowsheet (GF1) produces about 10 times as much nitrous oxide and ammonia and twice as much nitric oxide as the glycolic-formic flowsheet. The glycolic-formic flowsheet produces almost exclusively NO_2.

SRAT and SME Rheology

The rheology of the sludge simulant made testing very challenging. The initial sludge has a yield stress of 31.5 Pa and a consistency of 11.8 cP. This feed slurry exceeds the DWPF SRAT receipt design basis of 1-10 Pa. The high yield stress feed was expected to impact the yield stress values of the SRAT and SME products. During acidification in the SRAT, the slurry yield stress increases and reaches a maximum at approximately pH 7 (Figure 3-12). During all testing, this was a challenging processing segment and often required periodically turning the agitator off and back on to prevent the acids from pooling in the vessel. A software tool was added to the 4-L control computer to allow the automated turning off and turning on of agitation. In the 22-L vessels, a second agitator blade was added. In addition, the acid is fed below the liquid surface in the 22-L kettles so no pooling of acids was apparent.

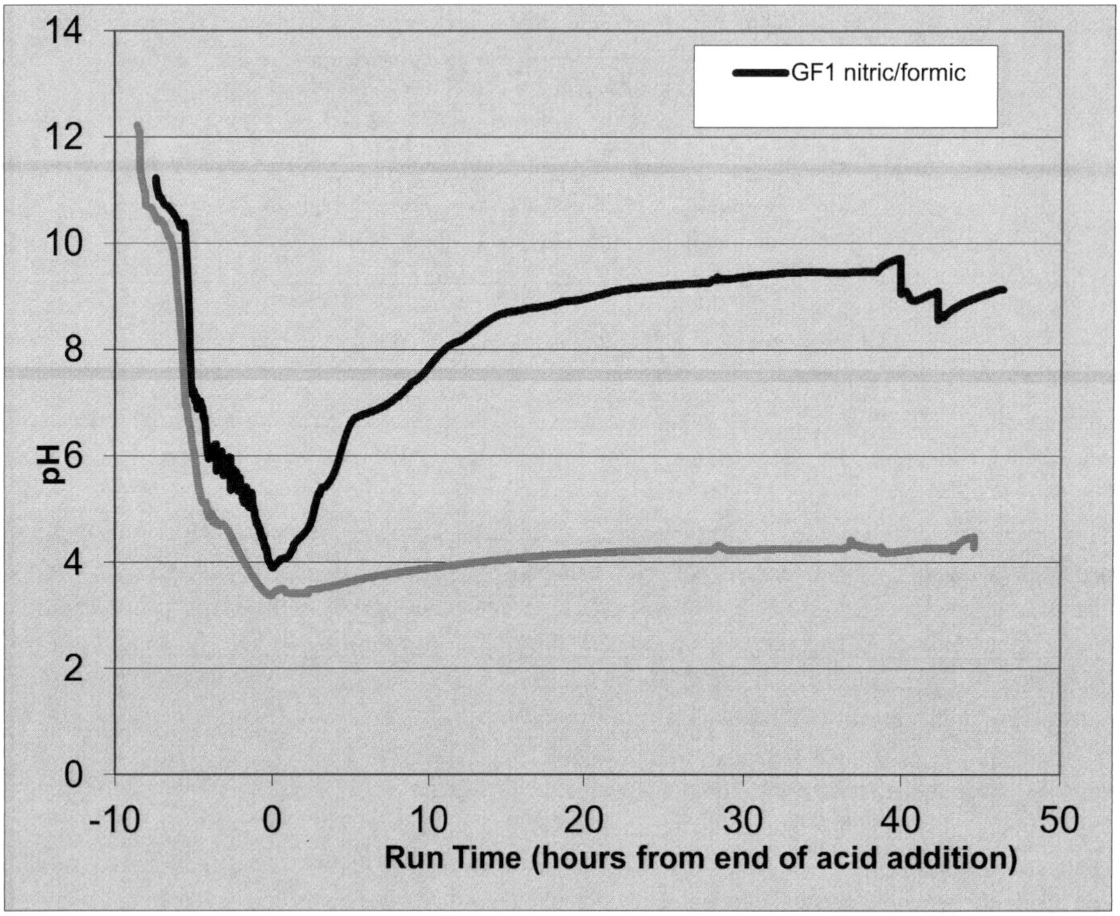

Figure **0-12**. pH trends for baseline SRAT/SME runs with mercury and noble metals

Once the pH of slurry dropped below 6, the slurry thinned considerably. In the baseline formic flowsheet, as was noted in Figure 3-12, the pH slowly increased during SRAT processing until it reached a pH of 9.5 at the end of the SRAT cycle. The SRAT product had a yield stress of 33 Pa and a consistency of 22 cP, exceeding DWPF SRAT design bases for both yield stress and consistency. The glycolic-formic flowsheet pH did not change appreciably after acid addition during SRAT processing. The SRAT product had a yield stress of 1.6 Pa and a consistency of 7.1 cP. These properties constitute a relatively low viscosity SRAT product. Flow curves for the SRAT and SME products were obtained by using a Haake RS600 rheometer and the current DWPF simulant rheology protocol[24]. The up and down curves (Figure 3-12) were fit to a Bingham plastic model to determine yield stress and consistency. Down flow curve data are the generally preferred choice for comparisons between systems. The data for all runs are tabulated below for the SRAT (Table 3-12) and SME (Table 3-13) along with a graphical comparison of the SRAT product yield stress (Figure 3-14) for the first seven runs.

[24] Koopman, D. C., Rheology Protocols for DWPF Samples, WSRC-RP-2004-00470, Savannah River Site, Aiken, SC, 29808 (October 2004).

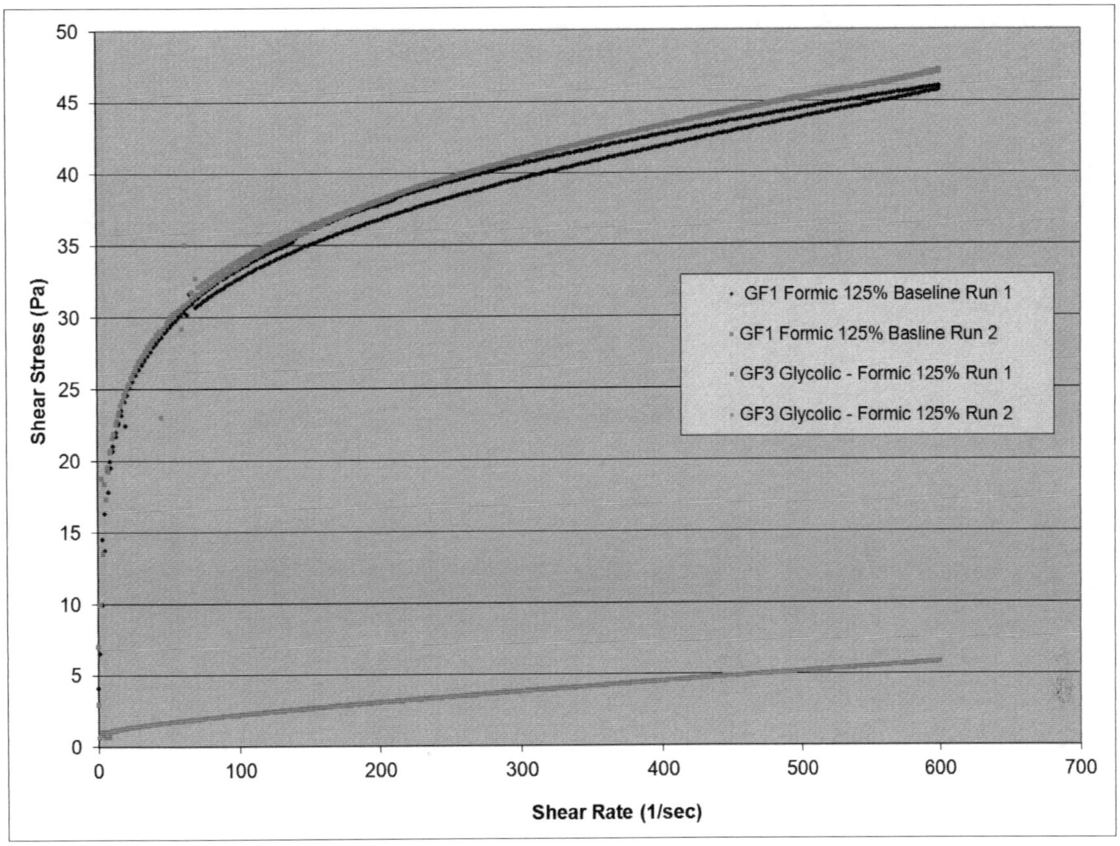

Figure **0-13**. SRAT Product flow curve comparison **for GF1 and GF3**

Table **0-12**. SRAT Product Rheology Summary

Rheology	GF1	GF2	GF3	GF4	GF5	GF6	GF7	GF8
Up yield stress, Pa	33.8	1.3	1.6	0.9	2.8	6.1	7.7	1.5
Up consistency, cP	22.2	6.6	7.1	5.0	10.1	14.8	10.3	6.4
Down yield stress, Pa	33.1	1.3	1.6	0.9	2.9	6.0	6.9	1.5
Down consistency, cP	22.8	6.6	7.1	5.0	10.0	14.7	11.0	6.4

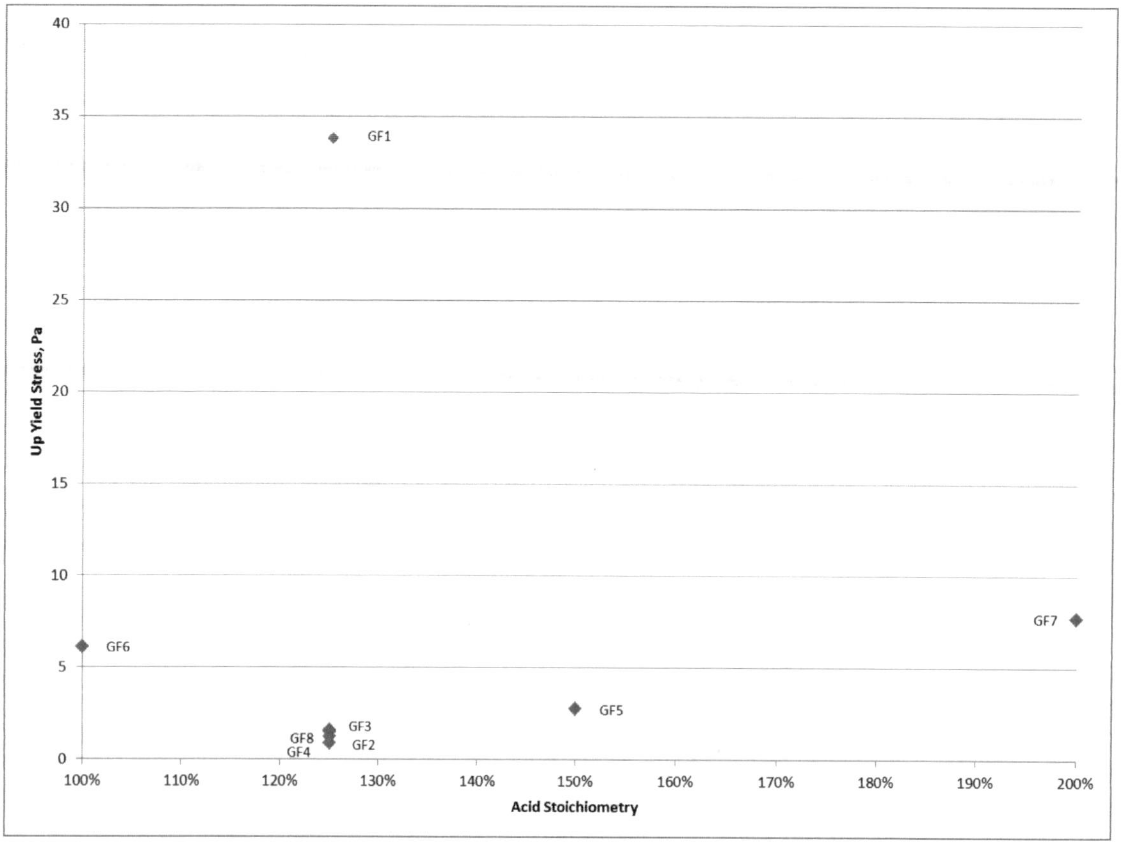

Figure **0-14**. SRAT Product Rheology

Table **0-13**. SME Product Rheology Summary

Rheology	GF1	GF2	GF3	GF4	GF5	GF6	GF7
Up yield stress, Pa	334.4	6.1	6.7	3.5	22.1	32.0	44.0
Up consistency, cP	-63.6	22.5	24.5	17.3	58.3	85.4	60.9
Down yield stress, Pa	223.0	6.8	7.2	3.3	23.9	30.2	41.4
Down consistency, cP	289.0	21.7	24.1	18.3	51.9	89.2	69.2

The lower yield stress was initially believed to be the result of the SRAT product differences of the two flowsheets. However, in producing melter feed for melt rate testing, shortened SRAT cycles were used to produce melter feed (runs GF23, baseline flowsheet, GF24, glycolic-formic flowsheet, GF25, sugar flowsheet). Each of these feeds was produced to target total solids of 45 wt% after the addition of frit. The final pH was 4.1 for the baseline flowsheet and 3.1 for the glycolic-formic flowsheet. In spite of similar pH profiles and SRAT product pH, the glycolic-formic flowsheet was much thinner rheologically than the baseline flowsheet. The yield stress of the glycolic-formic flowsheet is about one-third that of the other flowsheets. The rheology is summarized in Table 3-14. It is likely that the dissolution of metals, not pH is the major factor in the lower rheology for the glycolic-formic flowsheet. However, it should also be noted that there were higher yield stresses at higher acid stoichiometry.

Table **0-14**. Melter Feed Rheology Summary

Flowsheet	Baseline		Glycolic-formic		Sugar	Rheology
Run	**GF21**	**GF23**	**GF22**	**GF24**	**GF25**	**Limits**
Up yield stress, Pa	103	93	35	37	101	2.5-15
Up consistency, cP	44	41	43	27	38	10-40
Down yield stress, Pa	87	81	27	26	89	2.5-15
Down consistency, cP	62	56	49	40	55	10-40

The baseline nitric/formic flowsheet run far exceeds both the SRAT and SME design basis limits for yield stress and consistency. The formic/glycolic flowsheet run was also higher than limits but still fluid. The negative value for SME consistency in the baseline run indicates a large hump in the up curve; therefore the values for the down curve are more representative. Yield stress for GF7 is higher than that for GF5 or GF6 despite having more acid added and lower insoluble solids. This could be because the slurry chemistry is different at such a high acid stoichiometry or it could be sampling or analytical error. Additional rheology graphs are included in Appendix A2.

SRAT Acid Stoichiometry

Four glycolic-formic flowsheet runs were completed to develop an operating window for this flowsheet. Note that with the baseline formic flowsheet, no window was found. A wide acid stoichiometry window allows processing flexibility during DWPF operations. The stoichiometry window for the glycolic-formic flowsheet is at least from 100% to 200% of Koopman minimum acid demand. This is wider than any flowsheet that SRNL has developed for DWPF since sludge batch 1A (125-300). Generally the window for the formic acid flowsheet is limited by hydrogen and rheology, while the glycolic-formic flowsheet was not. The data are summarized in Table 3-15. Nitrite was successfully destroyed in all of these runs.

Table **0-15**. Acid Stoichiometry, Hydrogen Generation and Rheology

Run	Limit	GF6	GF3	GF5	GF7
Acid Stoichiometry	NA	100%	125%	150%	200%
SRAT Hydrogen, lb/hr DWPF Scale	0.65	0.013	0.006	0.008	0.024
SME Hydrogen, lb/hr DWPF Scale	0.23	0.012	0.004	0.013	0.049
SRAT product yield stress, Pa	1.5-5	6.1	1.6	2.9	7.3
SRAT product consistency, cP	5-12	14.8	7.1	10.3	10.6
SME product yield stress, Pa	2.5-15	31.1	6.9	23.0	42.5
SME product consistency, cP	10-40	87.3	24.3	55.1	65.1

SRAT/SME REDOX

SME product samples were vitrified in nepheline sealed crucibles and the resulting glasses measured for REDOX ($Fe^{2+}/\Sigma Fe$)[25]. The REDOX target for all the simulations in this study was 0.2 (except for runs GF13 to GF16 discussed below). The target is achieved by predicting the SME product anion concentrations and adjusting the split of acids between nitric and glycolic/formic. Therefore the ability to hit a REDOX target is highly dependent on being able to accurately predict anion behavior in the SRAT and SME cycles. Inserting the SME product data into the latest REDOX correlation gives a "predicted" REDOX that is different than the targeted REDOX of 0.2.

The REDOX prediction equation used in this study with an added term for glycolate is[26]:

$$Fe^{2+}/\Box Fe = 0.2358 + 0.1999 * (2[F] + 4[C] + 6\,[G] + 4[O_T] + 5[N] - 5[Mn])45/T$$

Where [F] = formate (mol/kg feed)
[C] = coal (carbon) (mol/kg feed)
$[O_T]$ = $oxalate_{Total}$ (soluble and insoluble) (mol/kg feed)
[G] = glycolate (mol/kg feed)
[N] = nitrate + nitrite (mol/kg feed)
[Mn] = manganese (mol/kg feed)

Values less than zero or greater than one can be calculated with the REDOX equation, because it is a linear regression equation fit to experimental data. Values outside the range of zero to one, however, are physically impossible. A number less than zero can be interpreted as fully oxidized and likewise a number greater than one as fully reduced.

Analysis of these data leads to two conclusions. First, predicted REDOX based on SME product data vary significantly from the targeted 0.2. Secondly, measured REDOX does not agree well with predicted REDOX.

The first conclusion points to errors in predicting anion conversion factors. This is not entirely unexpected for runs with a previously untested sludge and a new flowsheet. The results from these runs and in particular the 100% acid run can be used to generate new predictions for future work. It is easy to understand how it would be hard to predict REDOX if the reported anion concentrations vary substantially.

The second conclusion indicates error in either the applicability of the REDOX model (error in Mn or glycolate term) in this compositional space, the IC data, or the REDOX test method itself. The SRAT products were reanalyzed for anions (see Table 3-16). SRAT product samples were used since many of the runs were SRAT cycles only. The reanalysis of both the glycolate and nitrate were

---

25 Manual L29, Procedure ITS-0052, Rev. 2, Heat Treatment of Waste Slurries for REDOX (Fe2+/ΣFe) and Chemical Composition Measurement.

26 Jantzen, C. M. and M. E. Stone, Role of Manganese Reduction/Oxidation (REDOX) on Foaming and Melt Rate in High Level Waste Melters, WSRC-STI-2006-00066, Savannah River Site, Aiken, SC, 29808 , March 2007.

significantly different than the first analysis for many of the samples. The data was also used to predict the TOC concentration in the SRAT product, since no TOC analysis was requested of the SRAT or SME products.

Table **0-16**. SRAT product anions, mg/kg

Run	1st Nitrate	2nd Nitrate	1st Glycolate	2nd Glycolate	1st Formate	2nd Formate	1st Calculated TOC	2nd Calculated TOC
GF1	25,000	20,600	<100	<1000	12,900	12,250	3,520	3,350
GF2	51,600	54,500	53,900	68,500	2,555	2,495	18,300	23,100
GF3	55,950	52,650	54,700	56,400	<100	<1000	18,400	19,000
GF4	53,100	49,000	57,300	48,100	<100	<1000	19,700	16,700
GF5	59,900	68,800	115,000	85,500	824	<1000	38,300	28,700
GF6	45,250	51,500	66,300	51,300	776	<1000	22,600	17,800
GF7	59,900	71,650	161,000	130,000	3,750	3,245	53,700	43,500
GF8	56,250	62,100	118,000	75,700	<100	<1000	39,600	25,800
GF13	64,100	63,400	44,500	47,900	485	<1000	15,400	16,500
GF14	39,500	53,550	52,600	47,500	997	<1000	18,000	16,400
GF15	58,950	59,550	49,600	48,800	509	<1000	16,400	16,200
GF16	62,400	54,050	56,300	53,700	753	<1000	18,600	17,800
GF17	42,800	45,000	58,000	44,700	7,845	7,865	20,900	16,600
GF18	48,600	51,100	65,300	41,300	5,125	4,640	22,600	14,700
GF19	56,900	56,500	94,600	63,200	1,870	1,325	31,500	21,200
GF20	50,600	46,300	85,300	44,800	3,885	3,365	28,900	15,700

Table 3-17 below shows the appropriate SME product data with the corresponding predicted REDOX values as well as the REDOX as measured.

Table **0-17**. SME product data for REDOX calculations

anions (mg/kg)	GF1	GF2	GF3	GF4	GF5	GF6	GF7
Nitrite	<100	<100	<100	<100	<100	<100	<100
Nitrate	25,350	54,950	53,100	47,150	64,750	42,900	69,600
Glycolate	0	57,000	52,550	58,100	102,000	65,250	154,000
Oxalate	519.5	737	2690	4185	3500	4105	2175
Formate	22200	3470	<100	937	1700	2525	3455
Other properties							
Mn (gmol/kg slurry)	0.167	0.140	0.136	0.115	0.172	0.181	0.141
wt% total solids	46.28%	44.95%	44.66%	45.18%	49.73%	52.70%	47.04%
predicted REDOX*	0	0.159	0.108	0.336	0.654	0.432	1
measured REDOX	0.000	0.661	0.750	0.612	0.751	0.528	0.877
stoichiometry (% Koopman minimum)	125	125	125	125	150	100	200

* Predicted REDOX used existing REDOX prediction equation with Mn term

The initial formic-glycolic flowsheet runs (GF2 to GF7) all had very high glass REDOX results (0.5 to 0.8). This is much higher than the typical 0.2 target and outside the 0.1 to 0.33 REDOX limits. As a result, four runs (GF13 to GF16) were performed to target a glass REDOX of 0.0 to 0.3. The results are summarized in Table 3-18

Table **0-18**. Glass REDOX

Sample	GF13	GF15	GF14	GF16
REDOX Target, $Fe^{2+}/\sum Fe$	0.0	0.1	0.2	0.3
Predicted REDOX Calc without Mn term, $Fe^{2+}/\sum Fe$	0.03	0.08	0.16	0.22
Measured REDOX, $Fe^{2+}/\sum Fe$	0.00	0.08	0.20	0.45

In the REDOX series runs, the initial REDOX equation and estimates of anion conversion were used to develop the acid mix. Two issues had been noted. First the concentration of the anions is high so any error in the estimate of either glycolate or nitrate would skew that REDOX prediction. Second, the new glycolate method was being improved, and the PSAL analysts were getting more comfortable with the method. As a result, all of the anion analyses were repeated for the first twenty runs. However, even using new analyses, the REDOX equation prediction was much lower that the measured REDOX.

But if glycolic acid is completely reducing the Mn to +2 (the Mn is completely dissolved) and it stays reduced throughout the SRAT and SME processing, then there is no need for the Mn term.

A plot of measured REDOX against acid stoichiometry indicates that increasing acid led to linearly increasing REDOX in the test program, Figure 3-16 below. This strong correlation means that once the appropriate anion predictions for processing are determined, the REDOX should be predictable.

The ability to control REDOX once it can be accurately predicted depends on the quality of the SME product glycolate predictions and the amount of acid added. The less acid added, the smaller the effect conversion factor errors will have on the final REDOX.

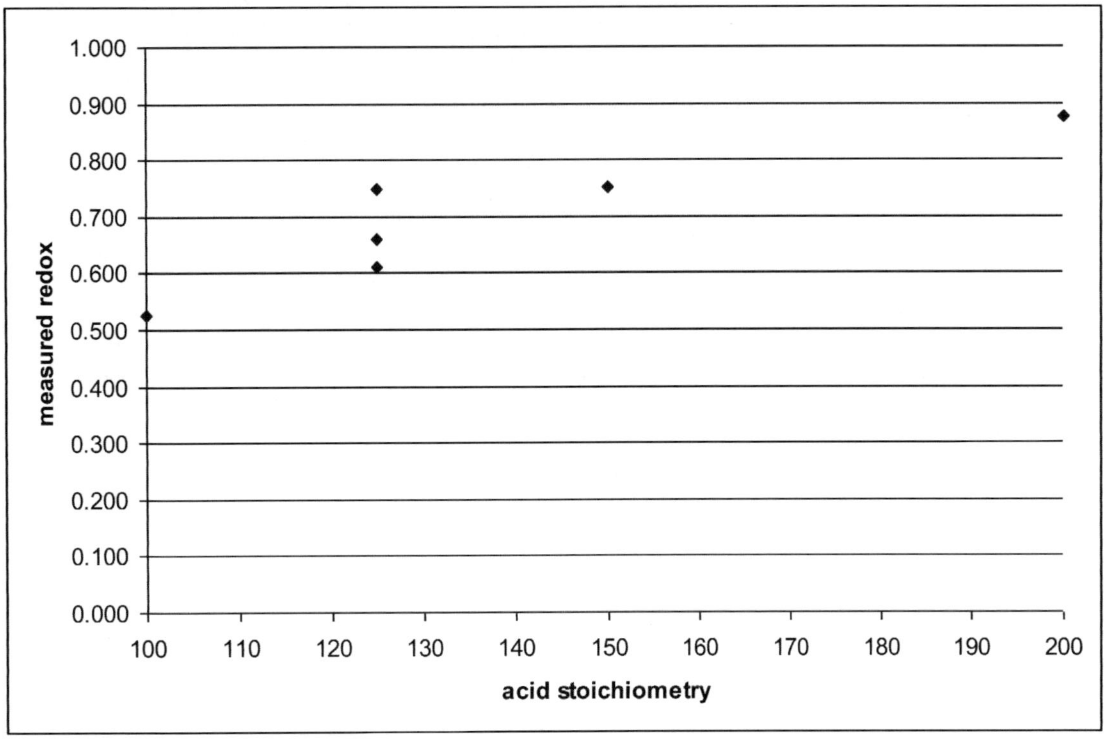

Figure **0-15**. Measured REDOX dependence on acid stoichiometry

One last note on REDOX is that the REDOX target does impact hydrogen generation. In the four runs targeting a REDOX of 0, 0.1, 0.2 and 0.3, the hydrogen generation increased with REDOX target. The run with the REDOX target of 0 had no detectable hydrogen. As the REDOX target increases, more reducing acid (glycolic-formic mixture) is added and less nitric acid is added. This leaves more formic acid present post acid addition for producing hydrogen. These runs all had low hydrogen generation as the runs were performed at a Koopman acid stoichiometry of 100%. The hydrogen profile for these runs is summarized in Figure 3-17.

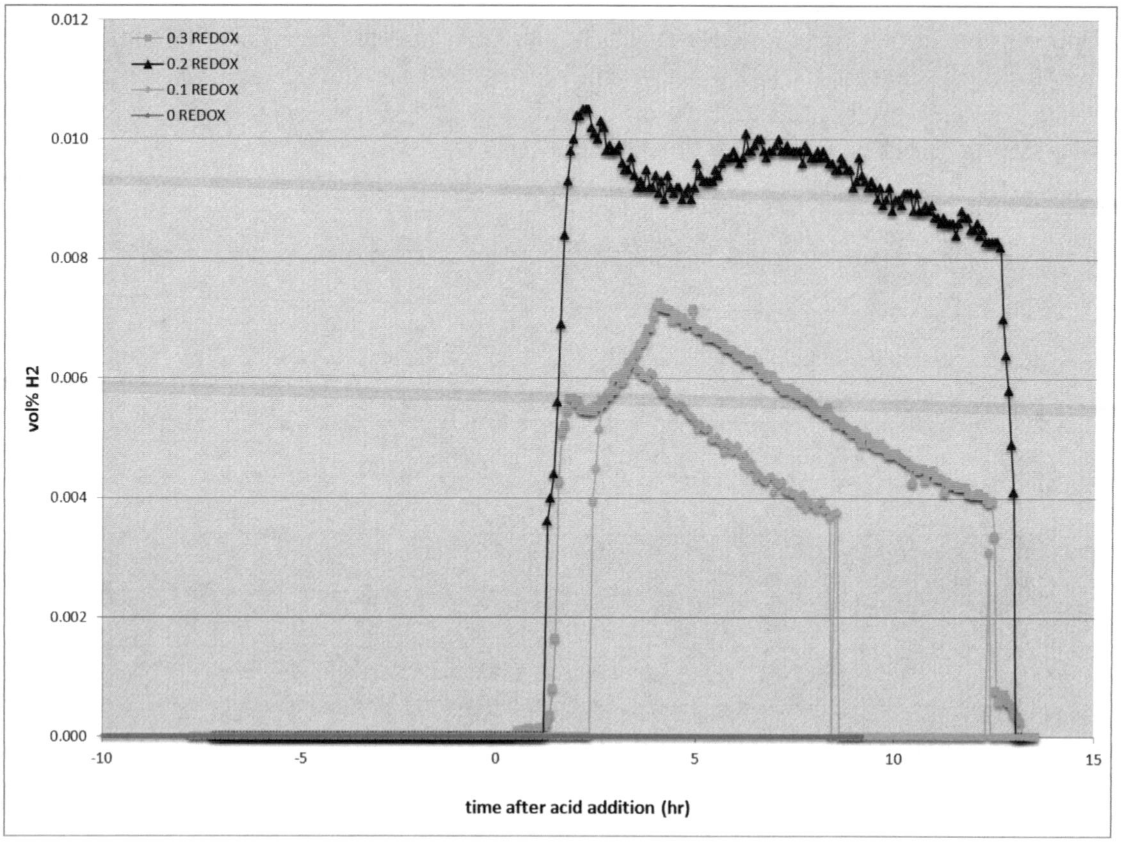

Figure 0-16. Hydrogen Generation Dependence on REDOX

An attempt was made to find an equation to better predict REDOX. One of the problems with this dataset is only one sludge was used. In addition, in targeting a REDOX of 0.2, the reducing anions (formate and glycolate) increased together with the oxidizing anions (primarily nitrate) as stoichiometry is increased. The REDOX equation assumes that the manganese is in a +5 oxidation state and must be reduced to +2 in the melter. But if glycolic acid is completely reducing the Mn to +2 (the Mn is completely dissolved) and it stays reduced throughout the SRAT and SME processing, there is no need for the Mn term.

Eliminating the Mn term increases the REDOX prediction by 0.1-0.2, giving the following revised REDOX equation.

$Fe^{2+}/\square Fe = 0.2358 + 0.1999 * (2[F] + 4[C] + 6 [G] + 4[O_T] +5[N])45/T$

Although the new Mn Free REDOX predictions agree better with the measured REDOX, it still can and should be improved. The results for all REDOX measurements are summarized in Figure 3-15. The three circled runs are the first three glycolic-formic runs (GF2 to GF4). The use of the REDOX data from the in-progress matrix study would be useful in improving the REDOX equation for the glycolic-formic flowsheet.

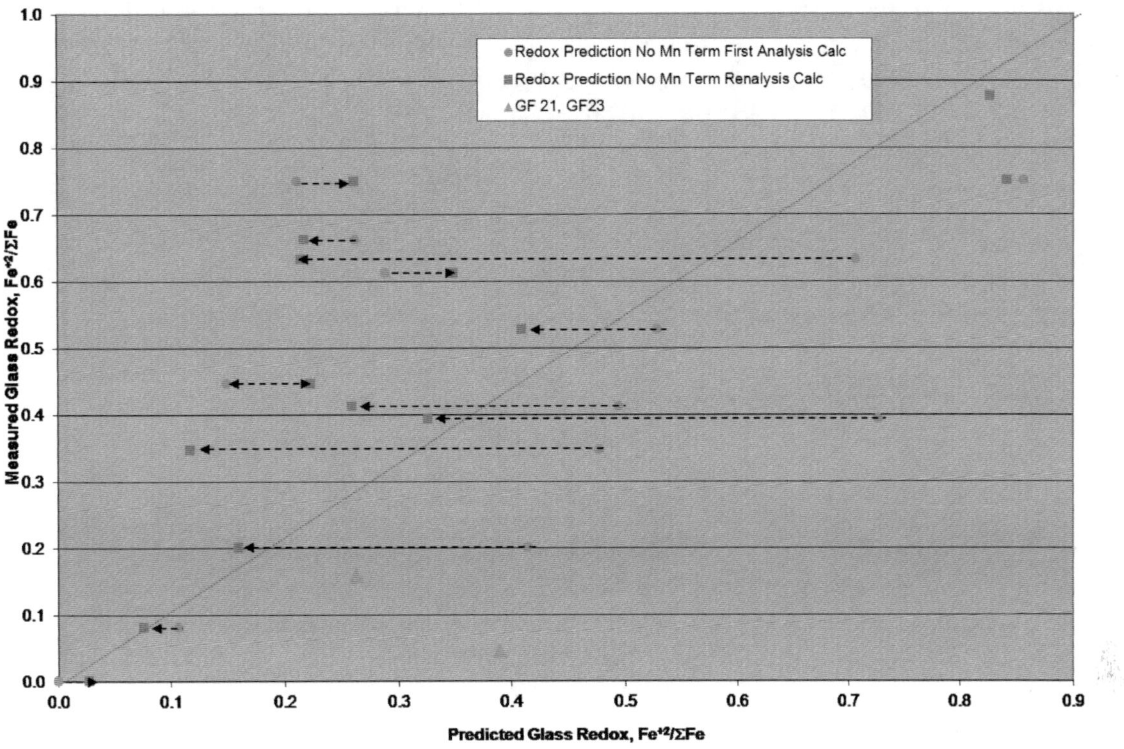

Figure **0-17**. SRAT Product REDOX comparison

<u>*SRAT Gd Solubility*</u>

Six SRAT runs were performed to compare the partitioning behavior of gadolinium to that of iron. This work was done to support potential use of gadolinium as a supplemental neutron poison[27]. Four of the SRAT runs were performed under typical baseline flowsheet conditions utilizing a combination of nitric and formic acids. The four baseline SRAT runs were SB6-27, -28, -29, and -30. In contrast, the other two SRAT runs were performed under alternative reductant flowsheet conditions utilizing a combination of nitric, formic, and glycolic acids. The two alternative reductant SRAT runs were GF-13 and -14.

In each of these runs, gadolinium was added to the sludge in order to determine the impact of the baseline flowsheet (SB6 runs) and the glycolic-formic flowsheet (GF Runs) on Gd solubility. The two GF runs were part of the REDOX series, GF13 having a REDOX target of 0 and GF14 having a REDOX target of 0.2. Both of these tests were at an acid stoichiometry of 100%. These tests are described in more detail in a report summarizing all of the Gd testing performed by Reboul[28]. For the baseline runs, the Gd and Fe became somewhat soluble (both peaked at about 20% soluble) and slowly declined to near zero by the end of the SRAT cycle (Figure 3-18). For the GF runs (Figure 3-19), the Fe

[27] S. H. Reboul, Task Technical and Quality Assurance Plan for Quantifying the Partitioning of Gadolinium in the Chemical Processing Cell, SRNL-RP-2010-00931, Savannah River National Laboratory, Aiken, SC, June 2010.

[28] S. H. Reboul, D. R. Best, M. E. Stone, D. R. Click, Partitioning of Gadolinium in the Chemical Processing Cell, , SRNL-STI-2010-00804, Savannah River National Laboratory, Aiken, SC, in draft.

remained insoluble but the Gd solubility continued to increase until almost 80% of the Gd was soluble. Thus Gd may not be as effective a neutron poison in the glycolic-formic flowsheet as it is in the baseline flowsheet, especially at higher acid stoichiometry..

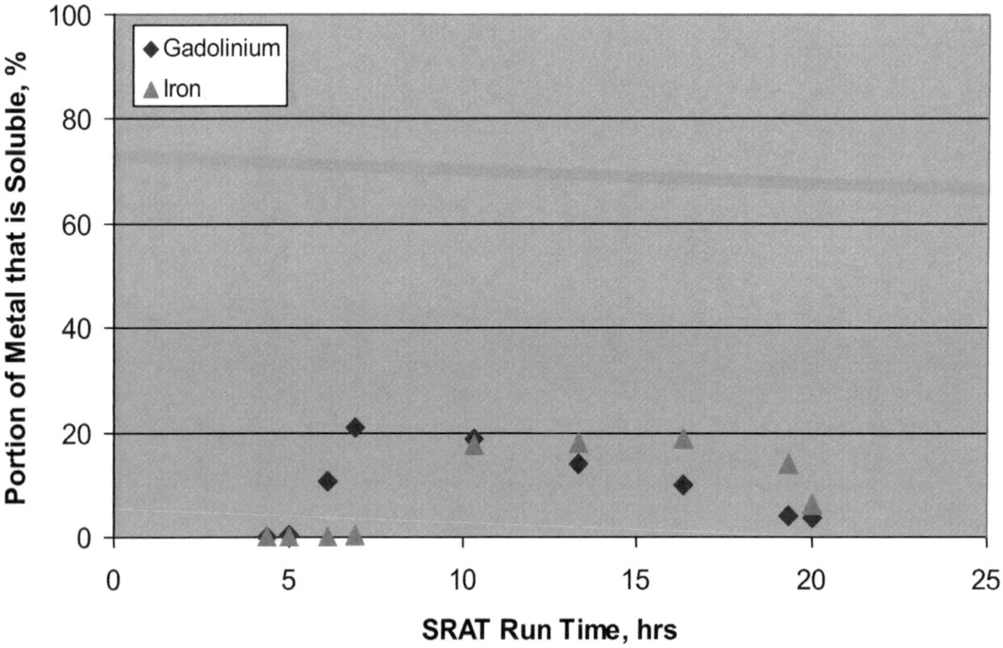

Figure **0-18**. SB6-29 Fe and Gd Solubility during SRAT Cycle

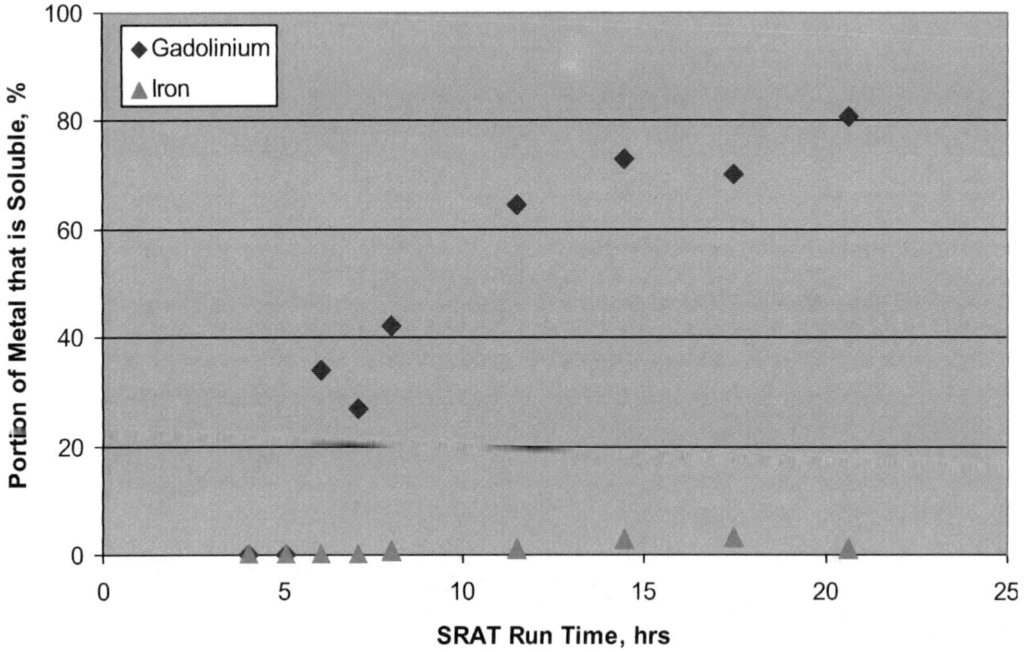

Figure **0-19**. GF13 Fe and Gd Solubility during SRAT Cycle

SRAT Mercury Reduction

One of the most important questions to resolve concerning the glycolic/formic flowsheet is whether mercury could be effectively reduced and steam stripped in the SRAT cycle. The starting sludge was trimmed to 3.263 wt% Hg in the total solids. (HM basis). This required a theoretical boiling time of nearly 36.5 hours to remove mercury to less than 0.60 wt% in the SRAT product total solids using lab-scaled DWPF design basis boil-up rates and a stripping efficiency of 750 g steam/g Hg. Note that DWPF has increased their mercury limit from 0.6 to 0.8 wt% in the SRAT product on a total solids basis.

Samples were taken periodically throughout the runs for mercury analysis. The chart below (Figure 3-20) shows the concentration of mercury in the slurry as a function of time for the two baseline flowsheet cases, GF1 (nitric/formic) and GF3 (nitric/glycolic/formic).

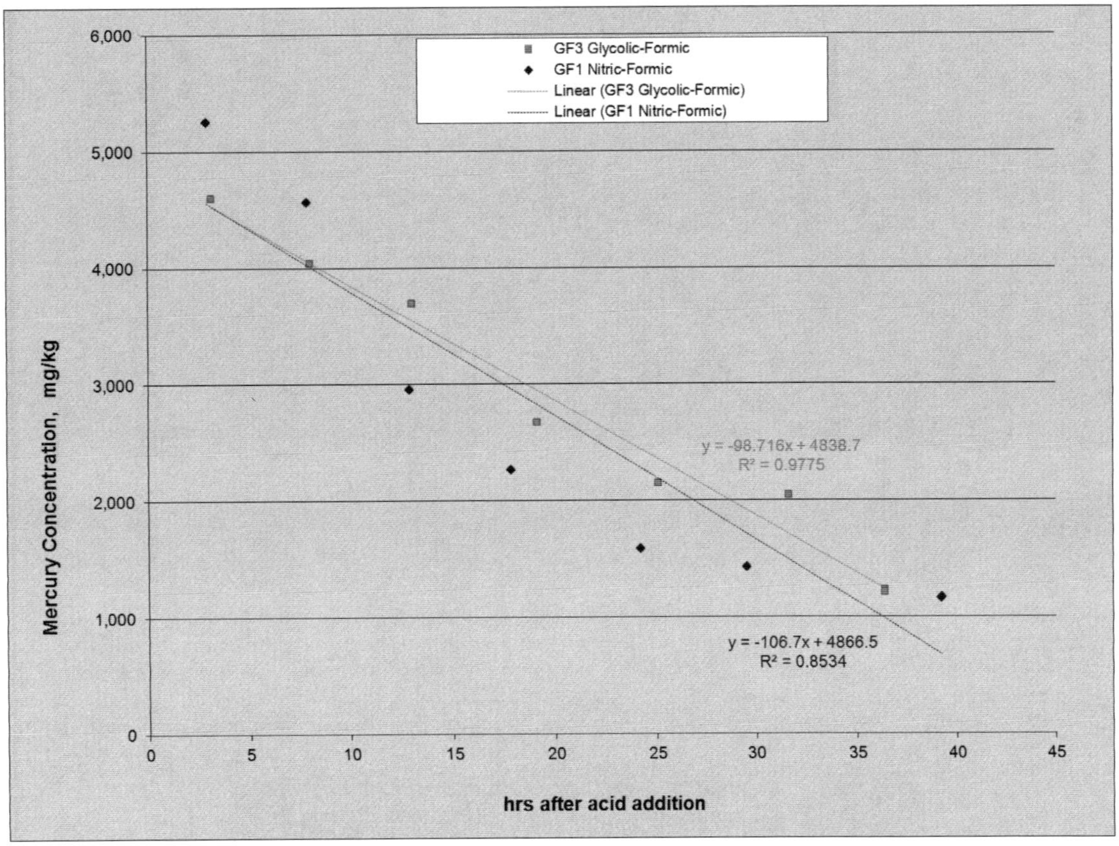

Figure **0-20**. Mercury concentration with time in baseline SRAT cycles

This plot shows actual sample data along with a calculated linear trend line for the two baseline flowsheets. The rate of mercury removal appears to be nearly identical for both flowsheets. On a total solids basis, the SRAT product Hg concentration in the glycolic/formic flowsheet was 0.56 wt% while the SRAT product for Hg concentration in the nitric/formic flowsheet measured 0.66 wt%. The next plot shows the mercury removal rate for the 100% stoichiometry run, GF6 (Figure 3-21). The rate of removal in this run GF6, the lowest acid stoichiometry tested, also matches well with the two baseline cases. The final SRAT product mercury concentration was slightly above the limit in this run, 0.61 wt% in the total solids. There are additional mercury graphs in appendix A1.

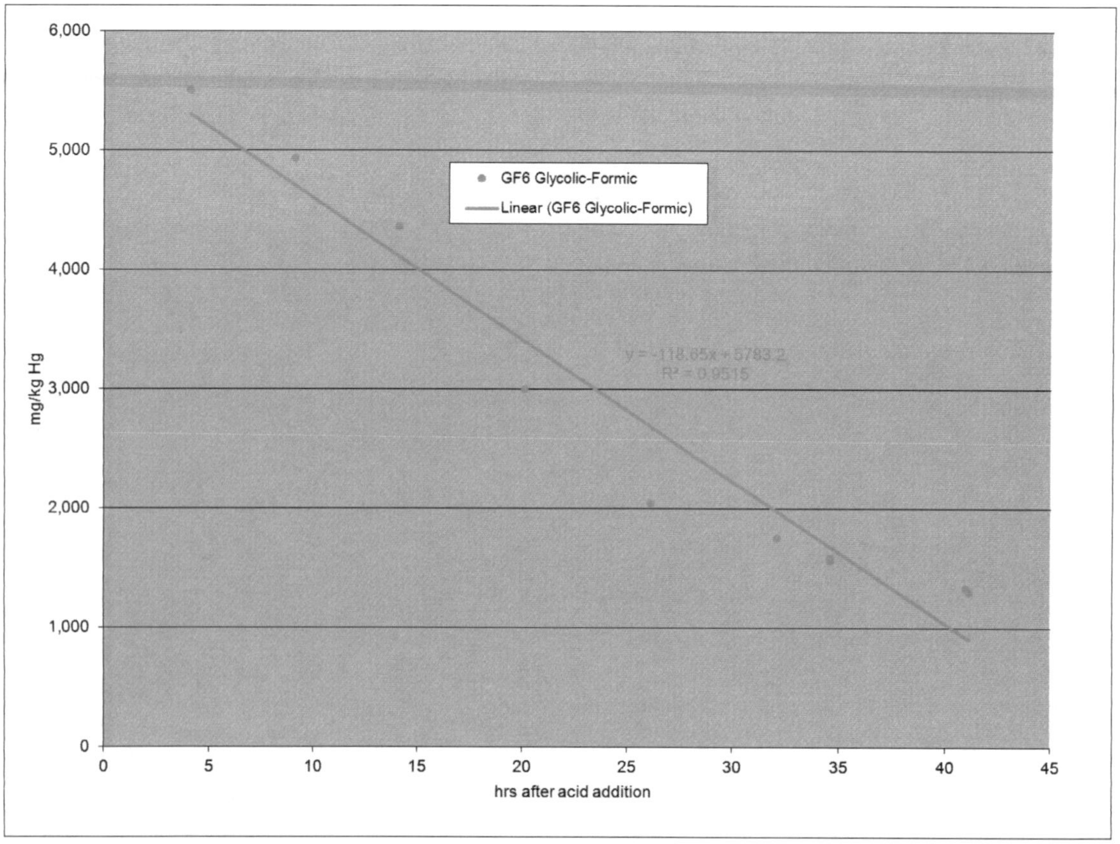

Figure **0-21**. Mercury concentration with time for 100% stoichiometry run

Table 3-19 below shows the SRAT and SME product mercury results for all of the runs in this series.

Table **0-19**. Comparison of SRAT and SME Product Hg results

Run	Acid Stoichiometry	SRAT Product Hg (wt% TS)	SME Product Hg (wt% TS)
GF1	125	0.66	0.29
GF3	125	0.56	0.25
GF4	125	0.97	0.39

GF5	150	0.41	0.15
GF6	100	0.61	0.25
GF7	200	0.31	0.37
GF8	125	0.39	no SME

All of the runs with the exception of the ARP/MCU run (GF4) had SRAT product mercury results near or below the limit of 0.60 wt% in the total solids. Note that in SRNL testing, the Hg was added after the ARP caustic addition was complete to prevent the formation of dimethyl mercury. In DWPF, some mercury is stripped during caustic boiling. However, since mercury is not added until caustic addition is complete; no mercury stripping was completed during caustic boiling.

It is of particular interest that mercury was successfully removed in GF8, a run with no formic acid. Glycolic acid did not show the ability to reduce mercury in earlier screening studies, which is why testing has focused on using the formic/glycolic blend. It is proposed that glycolic acid ($H_2(OH)CCO_2H$) is reacting with nitrite (or other species, e.g. MnO_2) to form glyoxylic acid ($HCOCO_2H$), the aldehyde form of glycolic acid. Glyoxylic acid then reacts with mercury oxide to form elemental mercury. The final form of the glycolic acid after these reactions is oxalic acid. This result potentially opens the door to completely removing formic acid from the flowsheet.

During the SRAT/SME process, mercury was collected in the Mercury Water Wash Tank (MWWT). The mercury tended not to coalesce into a single liquid phase, but instead come over as small beads. These beads would stick to the sides of the SRAT condenser drain tube rather than flowing to the sump in the MWWT. The mercury was dull in color. The sticking behavior qualitatively increased with increasing acid stoichiometry. The overall behavior of the mercury seemed to be consistent with previous observations of high acid runs in SB6 flowsheet studies.[29] One possibility is that the antifoam degradation is increased in higher acid runs (lower pH) and those degradation products form compounds with mercury, including calomel (Hg_2Cl_2). The photo below (Figure 3-22) was taken during the 200% acid stoichiometry run.

[29] Koopman, D.C., Sludge Batch 6/Tank 40 Simulant Chemical Process Cell Simulations, SRNL-STI-2010-00212, Savannah River National Laboratory, Aiken, SC, April 2010.

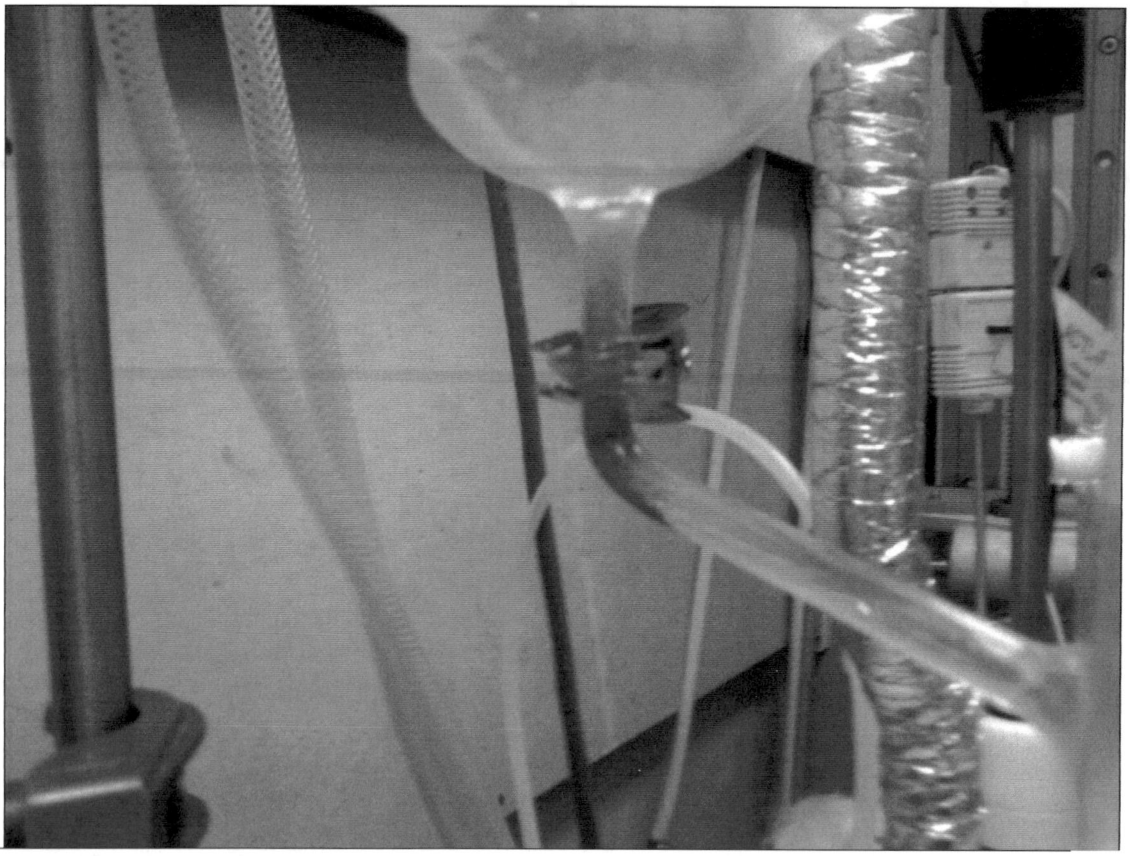

Figure 0-22. Mercury collection in SRAT condenser drain tube

All of the mercury removed and collected during these runs was retained for potential future analysis. Mercury stripping efficiency appears similar for the other runs. Plots of mercury concentration with time for all runs are included in Appendix A.

The mercury produced in the high acid runs was very different looking than the elemental liquid that was produced in the low acid runs. This is consistent with what is being produced in the baseline flowsheet runs for SB7. The mercury produced in the low acid runs is reduced and steam stripped easily, and it is easy to quantify the amount by weighing the mercury collected in the MWWT. The mercury produced in the high acid runs looks darker and more crystalline, collects in the tubing between the condenser and the MWWT, and mass-wise is much smaller than that collected in the low acid runs. Pictures of the mercury collected in the low stoichiometry and high stoichiometry runs are shown in Figure 3-23.

100% Acid Stoichiometry Silvery Bead Accumulation in Condenser to MWWT Leg	150% Acid Stoichiometry Minimal Dark Mercury Solids Accumulation in Condenser to MWWT Leg
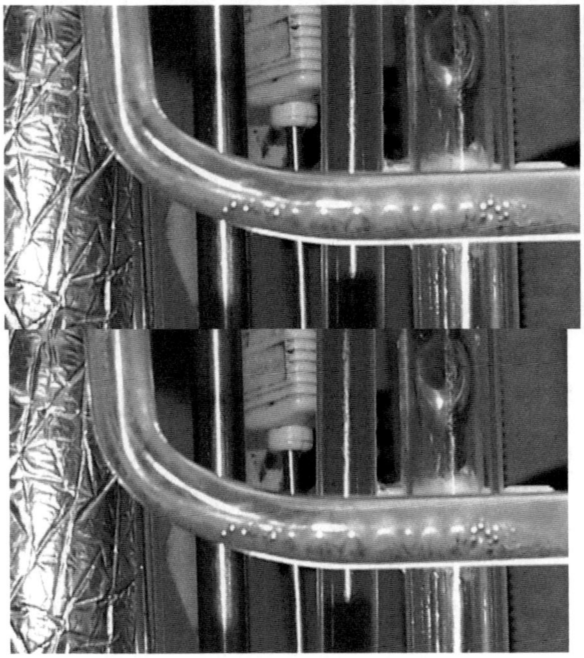	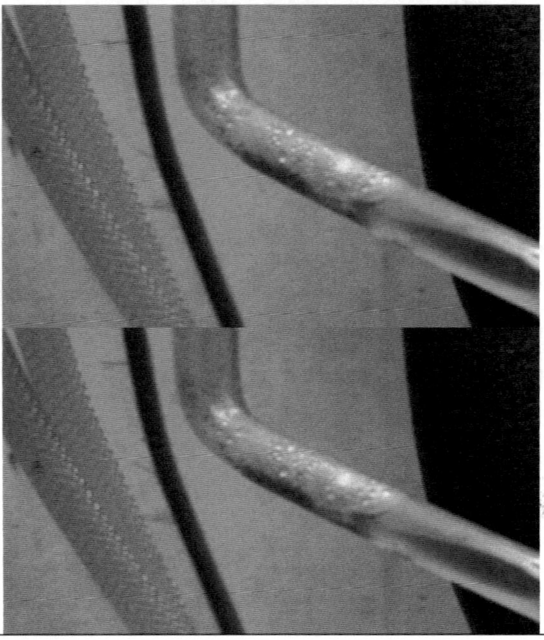

Figure **0-23**. Mercury collection in SRAT condenser drain tube

Because of the problems DWPF is experiencing with mercury recovery, a special study is being completed to understand the mercury mass balance. In several of the completed runs (GF17 to GF20), very careful cleaning of vessels with nitric acid and drying of the SRAT product to recover elemental mercury was completed. Table 3-202020 summarizes the data collected. At present the mass balance data from Run GF20 is complete and the data from Runs GF17-GF19 will be added when available. Small amounts of Hg were recovered in acid rinses of the glassware, but the major change was the recovery of about 45% of the mercury from the SRAT product that was not found in analysis of the SRAT product sample pulled from the SRAT vessel during cooldown.

Table 0-20. Mercury Balance

Run	GF17		GF18		GF19		GF20	
	Mass, g	%	Mass, g	%	Mass, g	%	Mass, g	%
Hg Added	NA	NA	NA	NA	NA	NA	12.12	100
SRAT Product Sample	NA	NA	NA	NA	NA	NA	1.75	14.4
SRAT Product Recovered	NA	NA	NA	NA	NA	NA	5.51	45.4
MWWT Hg	NA	NA	NA	NA	NA	NA	2.06	17.0
Condensate Hg	NA	NA	NA	NA	NA	NA	0.11	0.9
Total Hg Out	NA	NA	NA	NA	NA	NA	9.42	77.8

SRAT pH profile

Time dependent SRAT/SME pH data were collected throughout the first twenty-two runs. The graph below shows the pH trends of the two baseline runs with mercury and noble metals. The noise through the 5-7 pH region further illustrates the mixing difficulties experienced during acid addition. The data is summarized in Figure 3-24.

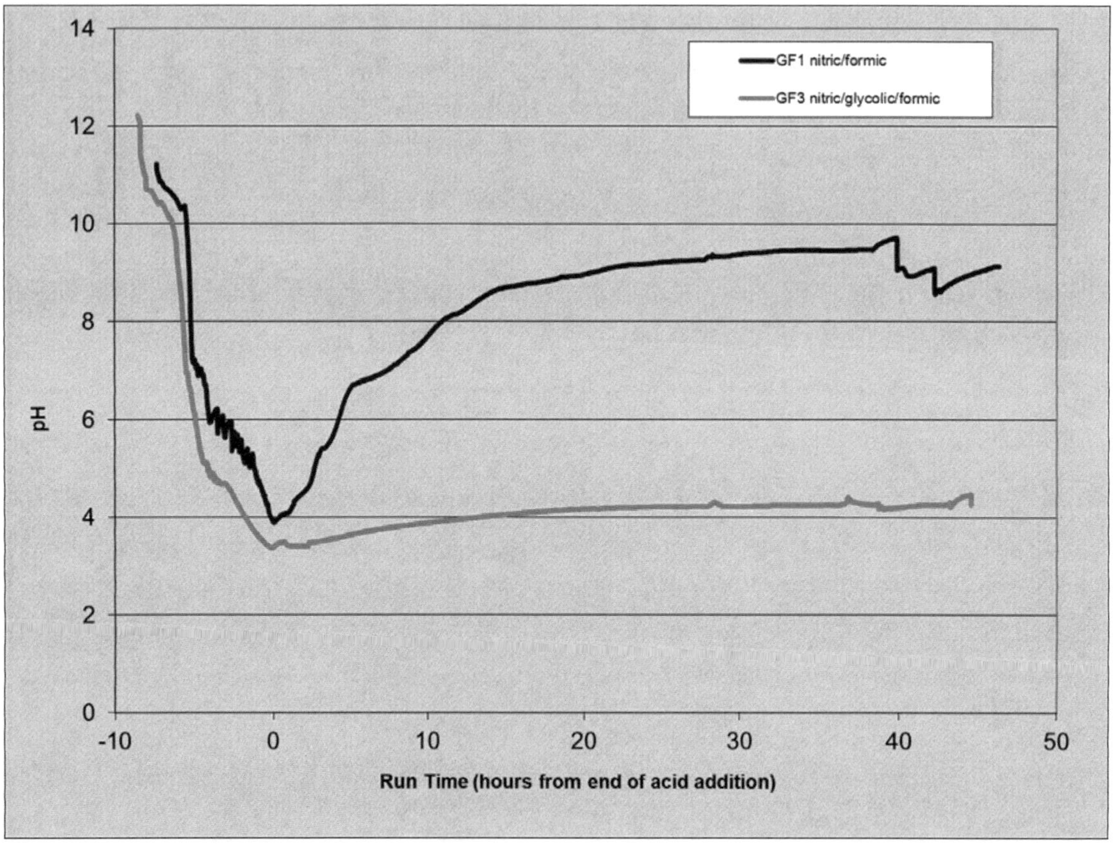

Figure **0-24**. pH trends for baseline SRAT runs with mercury and noble metals

The additional nitric acid needed for REDOX balancing the glycolic/formic acid flowsheet runs causes the pH to be lower at the end of acid addition. During reflux, the catalytic destruction of formic acid which causes hydrogen generation in the current DWPF flowsheet causes the pH to increase as formic acid is consumed. Since this does not happen to nearly the same extent in the glycolic/formic flowsheet, the pH tends not to rebound as much during reflux. This leads to low pH SRAT and SME products which presents some additional concerns related to metal solubility that will be discussed later. A positive result of the low pH with the alternate flowsheet is that yield stress both during the process and of the products is low compared to the baseline flowsheet.

The pH increase in the runs with noble metals and mercury is caused by the catalytic decomposition of formic acid. Note in two runs performed without mercury or noble metals (GF21 and GF22); there was no significant pH increase during the SRAT cycle for either flowsheet. The data is summarized in Figure 3-25.

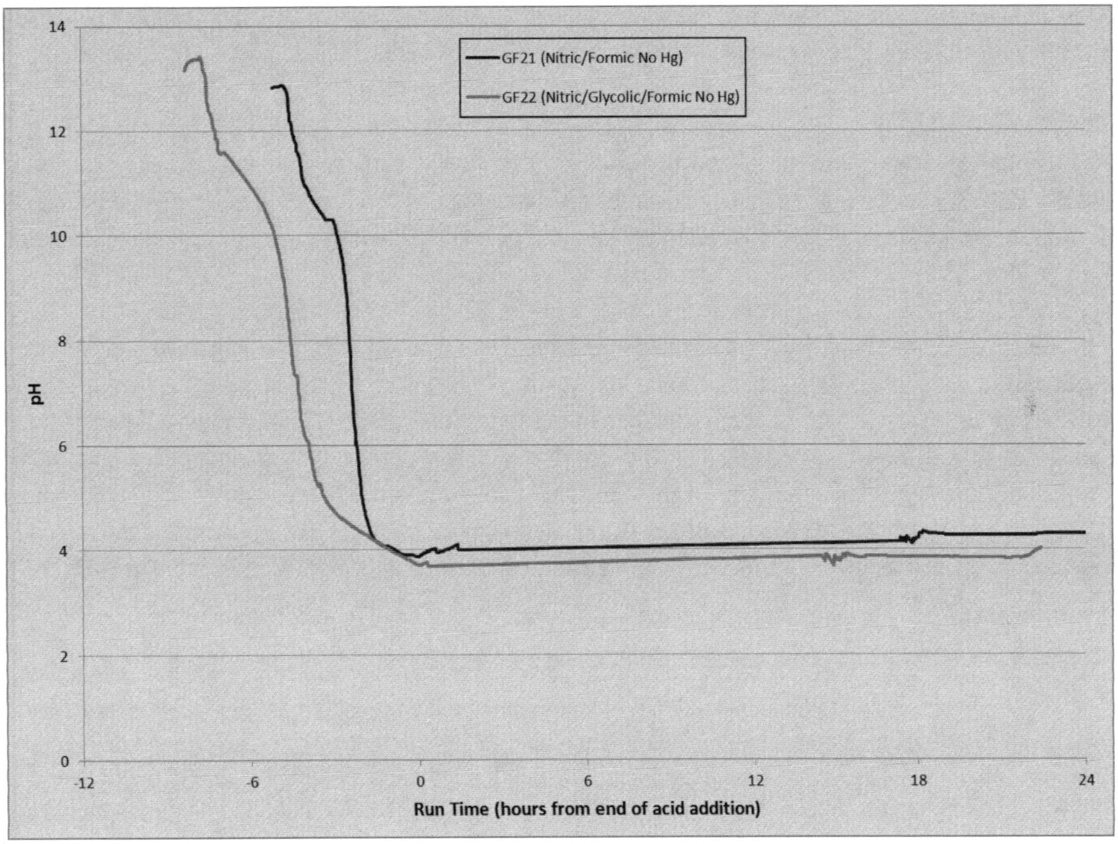

Figure **0-25**. pH trends for baseline SRAT runs without mercury and noble metals

<u>*SRAT Foaming Issues*</u>

One observation that could be made for all of the simulations is that maintaining a well-mixed vessel was very challenging with this particular sludge feed and was independent of the flowsheet testing objective. The sludge prepared at Harrell Industries measured 31.5 Pa yield stress. Because of the rheology of the sludge, it is possible that some of the problems noted in this testing would not be observed in actual DWPF processing, With a typical 4-L SRAT vessel, air entrainment becomes a serious concern during periods of significant off-gas generation, that is, from early in acid addition through the entire period of NO_x generation. A pocket of gas was found to build up around the agitator blades which prevented the bulk slurry from mixing. The agitator motor was stopped to allow the trapped gas to escape, followed by restarting the mixer. This method was effective for getting through acid addition without stopping acid flow, however, the mixer would need to be stopped and restarted at least once a minute. It is important to note that trapped air in the slurry caused by the unusually high yield stress is not the same as foaming in a classical sense. The trapped air caused an increase in slurry level, but no foam lamella was observed. Antifoam was added during one of the runs, but no decrease in slurry level was noted as a result of the addition. Again, this was common to all of the demonstrations with this simulant, not just the baseline run.

Another overall observation for these runs is that foaminess was not really a concern other than in the ARP/MCU test. The ARP/MCU run foamed extensively during ARP addition while caustic boiling and antifoam was added as necessary. Foaming during caustic boiling has been a minor issue with simulants, though it is known to be a problem with real waste.

SRAT Condensate Data

A composite sample was prepared for each of the first 8 runs by combining the condensate from the MWWT, FAVC, and SMECT to produce an "average" condensate sample. The sample was analyzed for elementals via ICP-AES, anions via IC, pH and density. The results are summarized in Tables 3-21 and 3-22.

The main difference in the flowsheets was the pH of the condensate samples and the amount of formate, glycolate and nitrate in the composite samples. In the glycolic-formic flowsheet runs, the condensate pH was 1-2.4. For the baseline formic-nitric flowsheet, the pH was 6.5. In the run with glycolic acid, no glycolate was measured in the condensate but was present in glycolic-formic flowsheet runs. One other observation is that the higher acid stoichiometry runs had higher Si in the condensate. This is likely due to antifoam degradation products in the condensate. It should also be noted that GF3 had almost 2X higher Si than GF1 indicating that the antifoam may degrade faster in the glycolic-formic flowsheet.

Table **0-21**. Comparison of Composite Condensate Data, ICP-AES, mg/L

Sample ID	GF1	GF2	GF3	GF4	GF5	GF6	GF7	GF8
Lab ID	10-0978	10-0979	10-0980	10-0981	10-0982	10-0983	10-0984	10-0985
Ag	<0.010	<0.010	<0.010	<0.010	<0.010	<0.010	<0.010	<0.010
Al	0.094	0.077	0.082	0.085	0.087	0.094	0.088	0.099
B	<0.100	<0.100	<0.100	<0.100	<0.100	<0.100	<0.100	<0.100
Ba	<0.010	<0.010	<0.010	<0.010	<0.010	<0.010	<0.010	<0.010
Ca	0.112	0.110	0.108	0.098	0.161	0.163	0.124	0.135
Cr	<0.010	<0.010	<0.010	<0.010	<0.010	<0.010	<0.010	<0.010
Cu	0.032	0.028	0.031	0.031	0.035	0.035	0.029	0.033
Fe	<0.010	<0.010	<0.010	<0.010	<0.010	<0.010	0.058	<0.010
K	0.262	0.211	0.223	0.215	0.243	0.302	0.212	0.208
Li	0.160	0.161	0.161	0.161	0.161	0.161	0.161	0.161
Mg	<0.010	<0.010	<0.010	<0.010	<0.010	<0.010	<0.010	<0.010
Mn	<0.010	<0.010	<0.010	<0.010	<0.010	<0.010	<0.010	<0.010
Na	<0.100	<0.100	<0.100	<0.100	<0.100	<0.100	<0.100	<0.100
Ni	<0.010	<0.010	<0.010	<0.010	<0.010	<0.010	<0.010	<0.010
P	0.095	0.085	0.096	0.047	0.082	0.096	0.103	0.078
Pb	<0.010	<0.010	<0.010	<0.010	<0.010	<0.010	<0.010	<0.010
S	<0.100	<0.100	<0.100	<0.100	<0.100	<0.100	<0.100	<0.100
Si	84.1	118.7	155.1	163.5	125.8	99.8	354.5	185.6
Ti	0.011	0.011	0.011	0.014	0.012	0.011	0.011	0.011
Zn	<0.010	<0.010	<0.010	<0.010	<0.010	<0.010	<0.010	<0.010
Zr	<0.010	<0.010	<0.010	<0.010	<0.010	<0.010	<0.010	

Table **0-22**. Comparison of Composite Condensate Data, IC (mg/L), Density (g/mL), pH

Sample ID	GF1	GF2	GF3	GF4	GF5	GF6	GF7	GF8
Lab ID	10-0978	10-0979	10-0980	10-0981	10-0982	10-0983	10-0984	10-0985
Fluoride	<100	<100	<100	<100	<100	<100	<100	<100
Chloride	<100	<100	<100	<100	<100	<100	<100	<100
Nitrite	<100	<100	<100	<100	<100	<100	<100	<100
Nitrate	156	224	132	132	1520	2230	4460	4180
Sulfate	<100	<100	<100	<100	<100	<100	<100	<100
Glycolate	<100	61.9	62.8	62.8	80.5	64.6	60.2	180
Oxalate	<100	<100	<100	<100	<100	<100	<100	<100
Formate	322	612	293	293	948	169	2310	<100
Phosphate	<100	<100	<100	<100	<100	<100	<100	<100
Density	0.998665	0.99878	0.998615	0.99882	0.999545	0.99974	1.00149	1.000885
pH	6.45	2.22	2.4	1.99	1.53	1.35	1.07	1.11

SME Offgas

Because of the long processing time in the SRAT cycle due to the stripping of mercury, there was no significant chemistry in the SME cycles (low carbon dioxide, hydrogen generation). Table 3-23 and Figure 3-26 summarizes the hydrogen and carbon dioxide generation in the SME cycle. The small generation of CO_2 and H_2 (Figure 3-26) is triggered by the addition of formic acid with the frit in the frit-slurry. This could be eliminated by not adding formic acid with frit in the SME cycle.

Table **0-23**. Comparison of SME Offgas Generation Data

SME Offgas	GF1 SME	GF3 SME
Nitrous Oxide (N_2O), g	0.00	0.00
Nitric Oxide (NO), g	0.00	0.00
Carbon Dioxide, g	3.6	1.7
Hydrogen, g	0.00	0.00

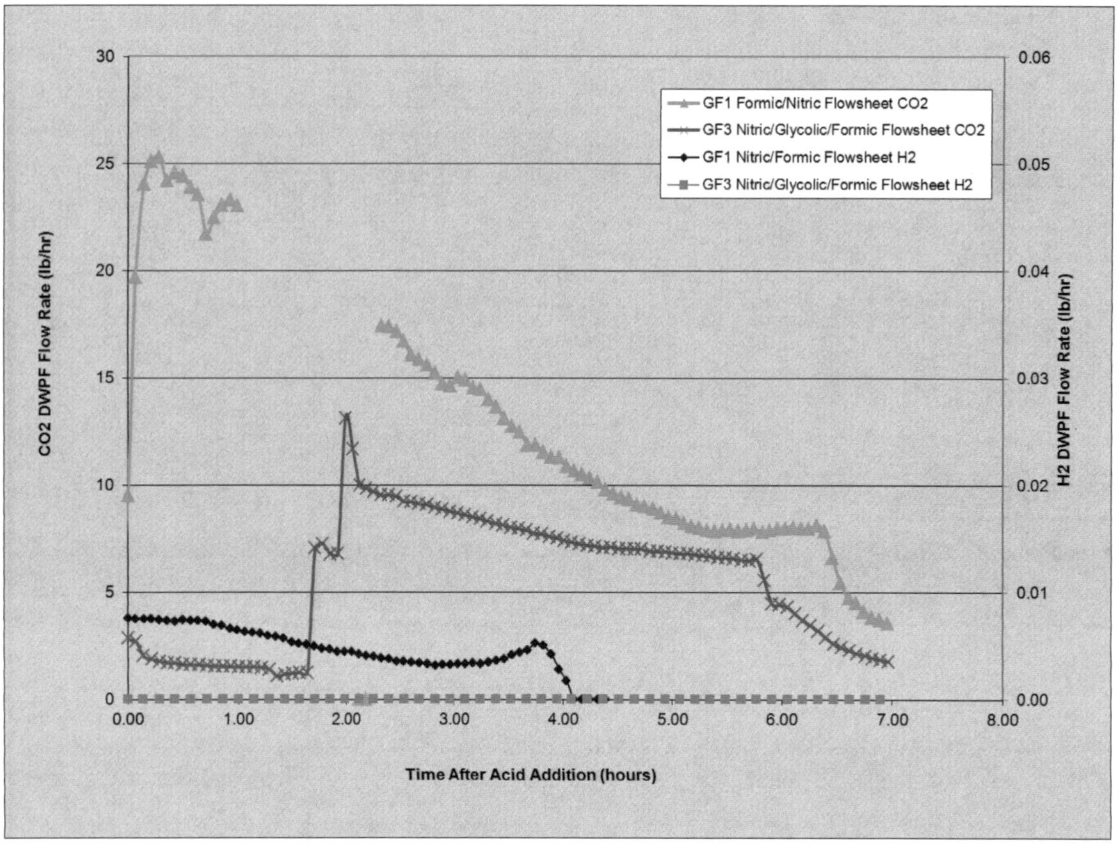

Figure **0-26**. Carbon Dioxide Generation in GF1 and GF3 SME cycles, lb/hr DWPF Scale

Ruthenium Testing

Four runs (GF9-GF12) were completed to understand the hydrogen generation in low and high acid stoichiometry runs with two forms of ruthenium. This was due to differences in hydrogen generation between the SRNL and VSL testing as the result of using different forms of Ru in the two labs. These tests are not pertinent to the flowsheet decision so will not be reported here.

Melter Feed Preparation

Two 4-L runs were completed to ensure the processing parameters were defined to produce a REDOX of 0.2 before the melter feed preparation began. This was needed since no runs without mercury, with 110% Koopman acid stoichiometry, had been completed. Run GF21 was the baseline flowsheet and Run GF22 was the glycolic-formic flowsheet run. A 4-L run with the sugar flowsheet was not completed as there was REDOX target for this run (sugar was added later to control REDOX). The conditions for these runs was then used as the inputs for the GF23 (baseline flowsheet) and GF24 (glycolic-formic flowsheet).

Three melter runs were performed at VSL to evaluate the performance of three flowsheets, baseline, glycolic-formic and sugar. The feed for these melter runs was prepared by SRNL in 22-L SRAT Simulant Equipment. Eight SRAT cycles were performed for each flowsheet to produce approximately 100 kg of SRAT product for melter testing. No frit was added at SRNL. The baseline flowsheet runs were designated as GF23A to GF23H. The glycolic-formic flowsheet runs were designated as GF24A to GF24H. The sugar flowsheet runs were designated as GF25A to GF25H.

<u>*Flowsheet Calculations of Acid Requirement*</u>

SRNL specified the conditions for the melter feed runs under the baseline and glycolic-formic flowsheets. VSL specified the conditions for the melter feed run for the sugar flowsheet. The run conditions are summarized in Table 3-24. The processing assumptions are summarized in Table 3-25, 3-26 and 3-27.

Table **0-24**. SRAT Cycle Additions and Dewater Targets, grams

Addition or Removal	GF23	GF24	GF25
Sludge Simulant	16,620	16,620	16,620
Other trim chemicals	5.7561	5.7561	5.7561
Water to dilute/rinse trim chemicals	500.00	500.00	500.00
Total Slurry	17,126	17,126	17,126
1:20 by mass Antifoam	266.48	266.48	266.48
50 wt% Nitric Acid solution	276.65	1,309.91	1,476.06
90 wt % Formic Acid solution	1,139.50	0.00	221.81
80:20 mole fraction Glycolic/Formic Acid solution	0.00	1,281.15	0.00
Water added to flush Acid Lines	20.00	20.00	20.00
Total Dewater	6,742.00	6,815.07	7,038.93

Table **0-25**. Sludge Composition for Melter Feed Preparation

Input	GF23, -25	Units
Fresh Sludge Mass without trim chemicals	16,620	g slurry
Fresh Sludge Weight % Total Solids	16.46	wt%
Fresh Sludge Weight % Calcined Solids	12.37	wt%
Fresh Sludge Weight % Insoluble Solids	8.45	wt%
Fresh Sludge Density	1.108	kg / L slurry
Fresh Sludge Supernate density	1.070	kg / L supernate
Fresh Sludge Nitrite	16,750	mg/kg slurry
Fresh Sludge Nitrate	6,135	mg/kg slurry
Fresh Sludge Formate	0	mg/kg slurry
Fresh Sludge Sulfate (mg/kg)	2,310	mg/kg slurry
Fresh Sludge Chloride (mg/kg)	776	mg/kg slurry
Fresh Sludge Phosphate (mg/kg)	0	mg/kg slurry
Fresh Sludge Oxalate	815	mg/kg slurry
Fresh Sludge Slurry TIC (treated as carbonate)	1,000	mg/kg slurry
Fresh Supernate TIC (treated as carbonate)	1,000	mg/L supernate
Fresh Sludge Hydroxide (Base Equivalents) pH = 7	0.643	Equiv Moles Base/L slurry
Fresh Sludge Coal/Carbon source	0.000	wt% dry basis
Fresh Sludge Manganese (% of Calcined Solids)	6.230	wt % calcined basis
Fresh Sludge Mercury (% of Total Solids in untrimmed sludge)	0.0000	wt% dry basis
Fresh Sludge Magnesium (% of Calcined Solids)	0.380	wt % calcined basis
Fresh Sludge Sodium (% of Calcined Solids)	19.500	wt % calcined basis
Fresh Sludge Potassium (% of Calcined Solids)	0.031	wt % calcined basis
Fresh Sludge Cesium (% of Calcined Solids)	0.000	wt % calcined basis
Fresh Sludge Calcium (% of Calcined Solids)	0.583	wt % calcined basis
Fresh Sludge Strontium (% of Calcined Solids)	0.000	wt % calcined basis
Fresh Sludge Nickel (% of Calcined Solids)	2.510	wt % calcined basis
Fresh Sludge Supernate manganese	0	mg/L supernate

Table **0-26**. SRAT Processing Assumptions for Melter Feed Preparation

SRAT Processing Assumptions	GF23	GF24	GF25	Units
mol % Glycolic Acid	0.00	80.00	0.00	mol glycolic acid/100 mol total acid
Conversion of Nitrite to Nitrate in SRAT Cycle	20.00	0.00	NA	gmol NO_3^-/100 gmol NO_2^-
Destruction of Nitrite in SRAT and SME cycle	100.00	100.00	NA	% of starting nitrite destroyed
Destruction of Formic acid charged in SRAT	20.00	90.00	NA	% formate converted to CO_2 etc.
Destruction of Glycolic acid charged in SRAT	0.00	0.00	NA	% glycolate converted to CO_2 etc.
Conversion of Glycolic acid to Oxalate	NA	3.00	NA	% glycolate converted to C_2O_4
Destruction of Oxalate charged	0.00	0.00	NA	% of total oxalate destroyed
Percent Acid in Excess Stoichiometric Ratio	110	110	72.7	%
SRAT Product Target Solids	27.84	29.39	28.31	%
Nitric Acid Molarity	9.444	10.005	10.005	Molar
Formic Acid Molarity	23.610	NA	23.385	Molar
Glycolic Acid Molarity	NA	11.826	NA	Molar
DWPF Nitric Acid addition Rate	4.6	4.6	4.7	gallons per minute
DWPF Formic Acid addition Rate	2.0	4.1	2.0	gallons per minute
REDOX Target	0.220	0.170	NA	Fe^{+2} / ΣFe
SRAT air purge	230	230	230	scfm

Table **0-27**. SME Processing Assumptions for Melter Feed Preparation

SME Processing Assumptions	GF23	GF24	GF25	Units
Sludge Oxide Contribution in SME (Waste Loading)	36.00	36.00	36.00	%
Target SME Solids total Wt%	45.0	45.0	45.0	wt%

Run Details

Eight duplicate 22-L batches of melter feed were completed to produce enough SRAT product for the VSL melter testing of each of the flowsheets. Processing conditions for these runs were modified to produce the batches as quickly as possible. For example, the acids were added at the same molar flowrate as formic acid so the nitric and glycolic-formic mixture were both added about 2x what is achieved in DWPF. Also, the SRAT boiling time was limited to the time needed to destroy nitrite and complete dewater to the target needed for added frit (36 wt % waste loading) to produce a 45 wt% total solids SME product. For the baseline and sugar flowsheet, this was too concentrated to feed to the melter so was diluted by VSL. Once the SRAT cycle was complete, the slurry was cooled down and transferred to the product drum. No cleaning of the processing equipment was completed between runs although the equipment was cleaned and reassembled before starting production of each flowsheet melter feed. Subsequent runs for each flowsheet were labeled with the run number and a letter designating which of the eight batches was produced. For example, Run 23A was the first run using the baseline flowsheet and Run 25H was the eighth run produced using the sugar flowsheet.

Product Results

The SRAT product from the first run for each flowsheet was analyzed, plus a composite sample from the drum was analyzed for each flowsheet. The composite results are summarized in Tables 3-28 to 3-32.

There are several analyses that are significant and will be discussed. First, the glycolic-formic flowsheet slurry is rheologically much thinner than the baseline or sugar flowsheet. Once the pH drops below 7 in the SRAT, the glycolic-formic flowsheet is much easier to mix. Second, the slurry soluble solid concentration is much higher for the glycolic-formic flowsheet. This is one of the reasons (pH is the other) for the lower yield stress for the glycolic-formic flowsheet. Third, the anions are very different for the three flowsheets. The baseline flowsheet is high in formate and nitrate. The formic-glycolic flowsheet is high in glycolate and nitrate. The sugar flowsheet is high in nitrate and will be high in sugar once added. Note that the sugar was added by VSL after all processing was complete and was not involved in SRAT or SME processing.

Table **0-28**. SRAT Product ICP-AES results, wt % calcined solids basis

Sample ID	10-GF-23-4501	10-GF-24-4519	10-GF-25-4512
Lab ID	11-0001	11-0002	11-0003
Al	16.1	15.75	15.95
Ba	<0.010	<0.010	<0.010
Ca	0.564	0.546	0.541
Cd	0.0455	<0.010	0.0580
Cr	0.122	0.125	0.127
Cu	0.269	0.250	0.269
Fe	20.9	20.65	20.4
K	0.0545	0.05	0.0545
Mg	0.450	0.445	0.449
Mn	6.29	6.21	6.23
Mo	<0.010	<0.010	<0.010
Na	16.5	16.2	16.3
Ni	2.69	2.64	2.69
P	<0.100	<0.100	<0.100
S	0.459	0.461	0.453
Si	0.157	0.158	0.173
Ti	0.0260	0.0250	0.0265
Zn	0.0120	0.0120	0.0135
Zr	0.252	0.254	0.262

Please note that the sugar flowsheet managed to get some nitrite destruction without any reducing agent although it did not meet the DWPF SRAT product specification. Consequently, it might be possible to get hydrogen generation in the SME cycle once the frit formic acid was added and the residual nitrite was destroyed. Also, the sulfate concentrations, as measured by the IC varied significantly for the three flowsheets. It would be expected that the sulfate concentration would be approximately equal in all three flowsheets. A row was added to Table 3-29, with a calculated sulfate concentration from the ICP-AES S analysis. As can be seen, the calculated sulfate concentrations were much more consistent than IC sulfate values.

Table **0-29**. SRAT Product Anion results, mg/kg

Sample ID	10-GF-23-4501	10-GF-24-4519	10-GF-25-4512
Lab ID	11-0001	11-0002	11-0003
Fluoride	<100	<100	<100
Chloride	1,090	515	965
Nitrite	<100	<100	1,170
Nitrate	27,050	68,600	79,200
Glycolate	<100	54,200	<100
Oxalate	<100	351	<100
Formate	71,400	9,520	12,700
Phosphate	<100	<100	<100
Sulfate	800	2,610	1,260
Sulfate from ICP S	2,380	2,170	2,300

Table **0-30**. SRAT Product Solids Results, wt % Total Solids Basis and pH

Sample	**GF-23-4501**	**GF-24-4519**	**GF-25-4512**
Lab ID	11-0001	11-0002	11-0003
Total Solids	27.1	28.4	26.6
Insoluble Solids	15.8	11.5	13.7
Calcined Solids	17.3	15.7	16.9
Supernate Solids	13.4	19.0	14.9
Soluble Solids	11.3	16.8	12.9
pH	4.2	3.0	4.7

Table 0-31. SRAT Product Rheology results

Sample ID	**GF23-4501**	**GF24-4518**	**GF25-4512**
Upcurve Yield Stress, Pa	93.5	36.9	101
Downcurve Yield Stress, Pa	80.6	25.8	89.0
Upcurve Consistency, cP	40.7	26.8	38.4
Downcurve Consistency, cP	56.2	39.7	55.1

The eight GF23 nitric/formic flowsheet runs were completed after first completing a 4-L run (GF21) to determine the processing parameters (REDOX being critical). The eight GF24 glycolic/formic flowsheet runs were completed after first completing a 4-L run (GF23) to determine the processing parameters. Results are summarized in Table 3-32. Note that runs GF23 and GF24 were shorter in duration than GF21 and GF22 as nitrite was <1000 mg/kg in four hours of boiling so the GF23 and GF24 were shortened to the time it took to dewater.

Table 0-32. SRAT Product REDOX, Inputs and Outputs

analyte	GF21	GF22	GF23	GF24
nitrite, mg/kg	<100	<100	<100	<100
nitrate, mg/kg	34,500	70,750	27,050	68,600
glycolate, mg/kg	<100	54,800	<100	54,200
oxalate, mg/kg	<100	802	<100	351
formate, mg/kg	68,400	9,550	71,400	9,520
wt% total solids	28.94	29.34	27.06	28.37
wt% calcined solids	18.22	16.25	17.27	15.69
Predicted REDOX, $Fe^{2+}/\Sigma Fe$	0.26	0.09	0.39	0.11
Measured REDOX, $Fe^{2+}/\Sigma Fe$	0.16	0.23	0.04	0.17

Note: REDOX was not measured for GF25, since it would have been completely oxidized prior to sugar addition (completed by VSL).

Recommended Flowsheet

The recommended Glycolic-Formic Flowsheet is very similar to the baseline flowsheet in most ways. The main change is to replace formic acid with an 80:20 molar blend of glycolic acid. It has been demonstrated that process was acceptable with all blends from 100% formic to 100% glycolic, so other acid blends might be preferable for some sludge batches. Glycolic acid is more dilute on a molar basis than formic acid (11.7 M for the mixture versus 23.6 M for 90 wt % formic acid), so it is recommended that glycolic acid is fed at the same molar flowrate as formic acid to keep from extending the SRAT processing time. Using the same basis for nitric acid would also be important as the nitric acid volume is larger in the glycolic-formic flowsheet so the acid feed time will be longer unless the acid addition rates are increased. The acid addition rate controls the production of process gases (CO_2, N_2O, NO_x, etc.) produced during acid addition.

The data in hand following the above studies indicate that the glycolic-formic acid flowsheet is technically feasible subject to additional progress in just one or two areas such as REDOX control. The existing formic acid cold feed preparation and handling equipment could be adapted for blend acid addition. Additional reduction in cycle time could be accomplished by increasing the nitric acid and formic acid feed pump capacity to four gallons per minute, but such modifications would not have to delay the implementation of the flow sheet. Few other facility modifications appear to be needed to implement this flowsheet in DWPF (some process instrumentation might need to be rearranged or modified), so the process could be deployed quickly once the necessary analytical methods, operating procedures, etc. were in place. Few downstream impacts have been identified. A low concentration of glycolate would be recycled to the tank farm (about a tenth of the formate concentration currently being recycled).

The acid stoichiometry needed will be specific to each sludge batch, but it is expected that a Koopman stoichiometry of 100% would be acceptable for most sludge batches. Higher acid stoichiometry might be needed if DWPF wants to target higher solids loading in the SRAT and SME or if the rheology of the sludge is challenging at lower acid stoichiometry. Hydrogen generation was minimal even at 200% acid stoichiometry and hydrogen generation can be virtually eliminated by using only glycolic acid. Lower acid stoichiometry led to more accumulation of elemental mercury in the MWWT.

The addition of glycolic acid in DWPF will require the analysis of the glycolate anion using ion chromatography. A new method has been developed for this analysis which allows the characterization of the current DWPF anions (nitrate, nitrite, formate, sulfate, phosphate, chloride, fluoride, and glycolate). Properties of the glycolic acid:formic acid blend fed to the SRAT will also need to be tracked in order to determine the correct volume to add.

Physical Properties

A document was issued summarizing the glycolic acid's physical properties, impurities, and radiation effects[30]. It concluded, "Blends of formic acid in glycolic acid were prepared and their physical properties tested. Increasing amounts of glycolic acid led to increases in blend density, viscosity and surface tension as compared to the 90 wt% formic acid that is currently used at DWPF. These increases are small, however, and are not expected to present any difficulties in terms of processing."

The main impurities in the glycolic acid are formic acid and sulfur. As long as it is being blended with formic acid, the formic impurity is not an issue. However, sulfur could impact glass quality. The report concluded, "The effect of sulfur impurities in technical grade glycolic acid was studied for its impact on DWPF glass quality. While the glycolic acid specification allows for more sulfate than the current formic acid specification, the ultimate impact is expected to be on the order of 0.03 wt% sulfur in glass. Note that lower sulfur content glycolic acid could likely be procured at some increased cost if deemed necessary."

Communication with DuPont after the issuing of the above report led to the discovery of DuPont data relating density and glycolic acid concentration at 50°C. This data was developed by titrating the acids, measuring the density and relating acid concentration to molarity. The data was used to calculate the concentration of the pure glycolic acid wt % and the resulting concentration of the glycolic/formic acid m Based on this data, the as purchased glycolic acid concentration was 71.1 wt% versus the 70 wt % assumed throughout the first 20 tests. The data is summarized in Table 3-33 below.

[30] Pickenheim, B.R. and Bibler, N.E, Glycolic Acid Physical Properties, Impurities, and Radiation Effects Assessment, SRNL-STI-2010-00314, Savannah River National Laboratory, Aiken, SC, May 2010.

Table **0-33**. Glycolic Acid Determination Using Density Measurement

Density, g/ml @50 ºC	wt% Glycolic
1.196373	60.7346
1.222427	67.3276
1.229959	69.0459
1.238992	71.2213
1.253999	75.3670
1.274718	80.2045
1.288263	83.4646

Degradation Issues

A document was issued summarizing glycolic acid's physical properties, impurities, and radiation effects[30]. It concluded "A paper study on the effects of radiation on glycolic acid was performed. The analysis indicates that substitution of glycolic acid for formic acid would not increase the radiolytic production rate of H_2 and cause an adverse effect in the SRAT or SME process. It has been cited that glycolic acid solutions that are depleted of O_2 when subjected to large radiation doses produced considerable quantities of a non-diffusive polymeric material. Considering a constant air purge is maintained in the SRAT and the solution is continuously mixed, oxygen depletion seems unlikely, however, if this polymer is formed in the SRAT solution, the rheology of the solution may be affected and pumping of the solution may be hindered." The proposed test is in progress and will be reported when complete.

Analytical Issues

As was discussed in section 2.3, a new analytical method was developed for the glycolic-formic flowsheet. The method is has a long analytical time due to the current necessity to analyze for phosphate. Measuring phosphorus by ICP-AES and remove phosphate from the list of requested anions could decrease analytical time significantly. The method is also sensitive to the potential degradation of standard solutions so it is recommended that all organic anion standards (manufacturer and calibration) be kept refrigerated and in the dark. Analytical QC will be needed on the glycolic acid-formic acid blend in the SRAT acid feed tank so that the molarity and acid proportions are known in order to determine the acid volume and acid mix to control REDOX.

Unresolved Technical Issues

The primary unresolved technical issue is the inability to predict REDOX based on anion measurements. This could lead to a new REDOX model for the glycolic-formic flowsheet or could be caused by inaccuracies in measuring anion concentrations. One of the hindrances to predicting the REDOX was the use of the same sludge for all testing. At present, a series of eight runs are being completed to study the glycolic-formic flowsheet using a matrix of four sludges (Hi Fe, Hi Al, Hi Mn, and Lo Mn). The variability in the composition of the sludge should help to refine or improve the REDOX model for the glycolic-formic flowsheet. The four REDOX runs (GF13, GF14, GF15, and GF16) demonstrated that melter feed could be produced to target the expected REDOX range of 0 to 0.35.

Conclusions

Flowsheet testing was performed to develop the nitric/glycolic/formic acid flowsheet as an alternative to the nitric/formic flowsheet currently being processed at the DWPF. This new flowsheet has shown that mercury can be removed in the Sludge Receipt and Adjustment Tank (SRAT) with minimal hydrogen generation. All other processing objectives were also met, including greatly reducing the Slurry Mix Evaporator (SME) product yield stress as compared to the baseline nitric/formic flowsheet. Forty-six runs were performed in total, including the baseline run and the melter feed preparation runs. Significant results are summarized below:

Constraint	Limit	Baseline Flowsheet GF1	Glycolic-formic Flowsheet GF3
SRAT hydrogen, lb/hr	<0.65	1.62	0.03
SME hydrogen, lb/hr	<0.23	0.0072	0.0017
SRAT carbon dioxide, lb/hr	NA	375	200
SRAT nitrous oxide, lb/hr	NA	0.75	1.93
SRAT product Hg, wt %.	0.8	0.66	0.56
SRAT product nitrite, mg/kg	<1000	<100	<100
SRAT product down yield stress, Pa	1.5 to 5	33.1	1.6
SRAT product down consistency, cP	5 to 12	22.8	7.1
SME product down yield stress, Pa	2.5 to 15	223	7.2
SME product down consistency, cP	10 to 40	289	24.1
Glass REDOX $Fe^{+2}/\Sigma Fe$	0.1-0.33	0.00	0.22[31]
SME product total solids, wt %	>45	46.28	44.66
Minimal foaming	NA	No	Yes

The baseline nitric/formic flowsheet run was extremely difficult to process successfully under existing DWPF acceptance criteria with this simulant at the HM levels of noble metals. Certain baseline results were not considered typical of normal flowsheet processing. While nitrite was destroyed and mercury was removed to near the DWPF limit, the rheology of the SRAT and SME products are well above design basis and hydrogen generation far exceeded the DWPF limit. In addition, mixing during the SME cycle was very poor. In this sense, the nitric/glycolic/formic acid flowsheet represents a significant upgrade over the current flowsheet. Mercury was successfully removed with almost no hydrogen generation and the SRAT and SME products yield stresses were within process limits or previously processed ranges.

The glycolic-formic flowsheet has a very wide processing window. Testing was completed from 100% to 200% of acid stoichiometry and using glycolic-formic ratios from 40% to 100% glycolic acid. The testing met all processing requirements throughout these processing windows. This should allow

[31] REDOX of GF14.

processing at and acid stoichiometry of 100% and a glycolic-formic ratio of 80% glycolic acid with minimal hydrogen generation. It should also allow processing endpoints in the SRAT and SME at significantly higher total solids content. The flowsheet should be easy to implement using the existing formic acid equipment with at most minor modifications. A modified IC method will be needed in DWPF to quantify glycolate in the SRAT and SME product samples. Such a method has been developed and implemented in SRNL.

Recommendations

The glycolic/formic flowsheet is recommended as a viable flowsheet alternative to the baseline DWPF flowsheet. In the testing that has been performed to date, this flowsheet meets or outperforms the current flowsheet in minimizing off-gas generation, mercury removal, thinner product rheology and having a wide processing window regarding both the glycolic-formic ratio and acid stoichiometry. The addition of glycolic acid leads to SRAT and SME products that are rheologically much thinner which means that more concentrated products can be produced, leading to potentially larger batches and higher throughput. Recommended SRAT processing conditions are summarized below:

- Hsu Acid Stoichiometry: 107.7% (100% Koopman Stoichiometry)
- Nitric Acid Concentration: 10.4 M (50 wt %)
- Nitric Acid Addition Rate: 4.5 gallon per minute (178.7 mole/minute)
- Glycolic/Formic Acid Concentration: 13.1 M (73.1 wt %)
- Glycolic-Formic Acid Addition Rate: 3.6 gallon per minute (178.7 mole/minute)
- 4 mols glycolic acid:mol formic acid
- Glycolic-formic Acid Density: 1.2576 g/mL at 50° C
- If needed, add glycolic acid in place of formic acid in frit slurry
- Targeting higher solids loading in SRAT and SME products

Two tests in progress should be completed and the results documented to provide additional data which may be important to making a decision on the best flowsheet for DWPF.

- Complete irradiation testing to determine whether irradiation of SRAT simulant supernate causes polymerization of the glycolic acid. Introducing glycolic acid to high radiation doses may cause some polymerization of the glycolic acid in an oxygen depleted atmosphere. Results will be reported when complete.
- Complete glycolic-formic flowsheet testing of the four matrix sludges to demonstrate the viability to process the very different sludge compositions of these four extreme simulants. This knowledge will provide additional confidence of the robustness of the glycolic-formic flowsheet if it is able to successfully process four sludge extremes planned in this testing matrix.

Based on the additional testing above, the REDOX equation should be modified or a new REDOX equation should be developed for predicting the melter REDOX for the glycolic-formic flowsheet.

The glycolate IC method could benefit from further refinement. Shortening the method cycle time and an overall improvement of accuracy would be very beneficial for both continued SRNL testing and eventual DWPF implementation. The run time is significantly long due to the request for phosphate. Measuring phosphorus by ICP-AES and remove phosphate from the list of requested anions could decrease analytical time significantly.

The glycolic-formic flowsheet should be demonstrated with actual waste in SRNL Shielded Cells SRAT and SME processing, including the production of glass and measurement of the glass REDOX.

Appendix A

A.1 Mercury Removal Plots

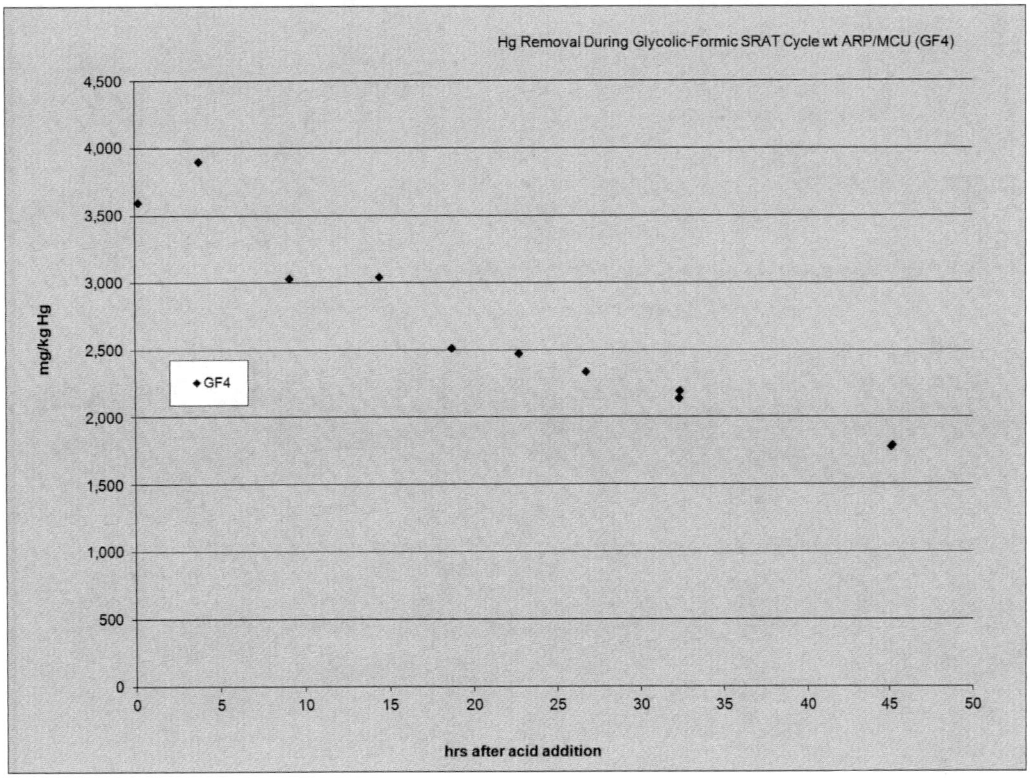

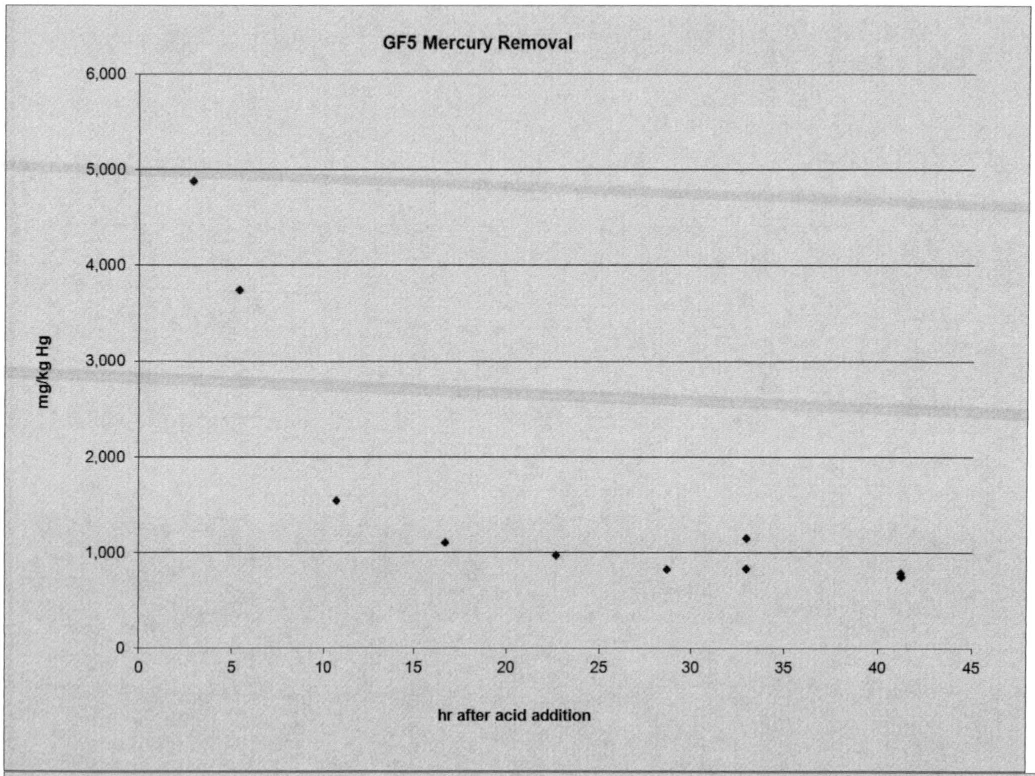
GF5 Mercury Removal
6,000
5,000
4,000
3,000
2,000
1,000
0
mg/kg Hg
0
5
10
15
20
25
30
35
40
45
hr after acid addition

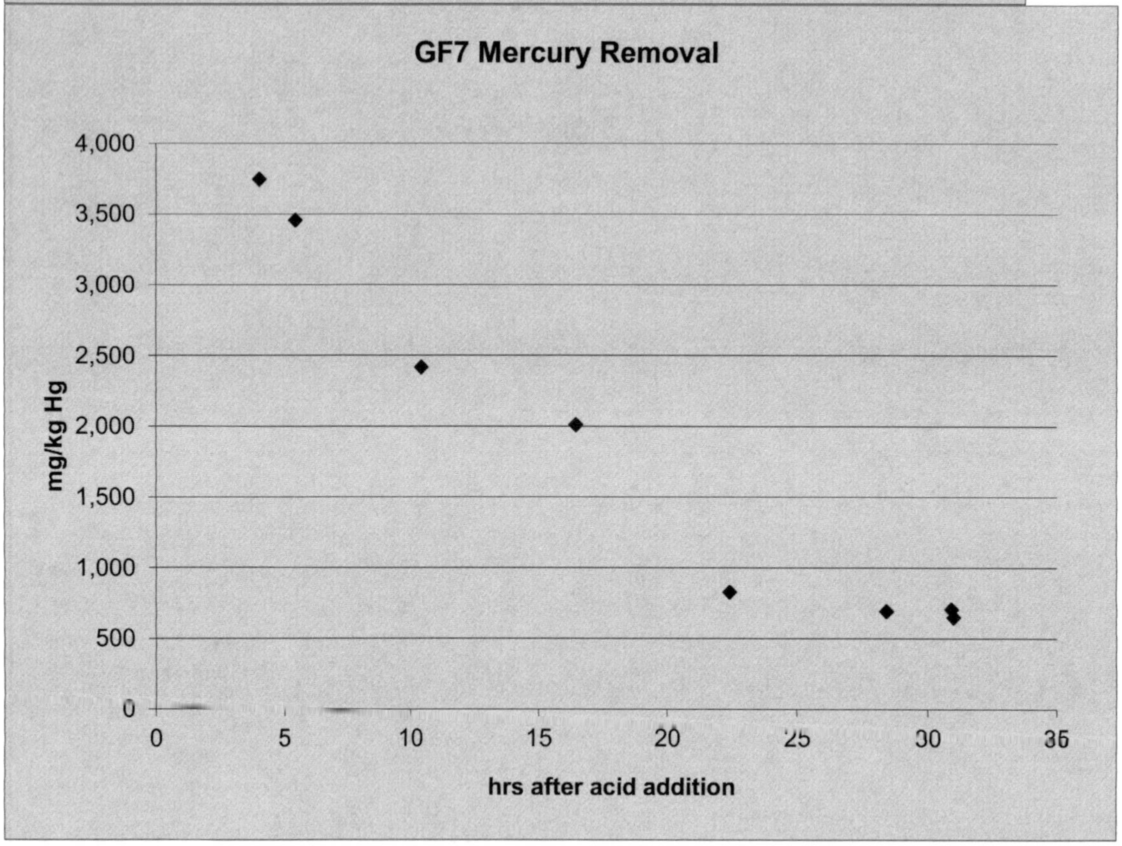
GF7 Mercury Removal
4,000
3,500
3,000
2,500
2,000
1,500
1,000
500
0
mg/kg Hg
0
5
10
15
20
25
30
35
hrs after acid addition

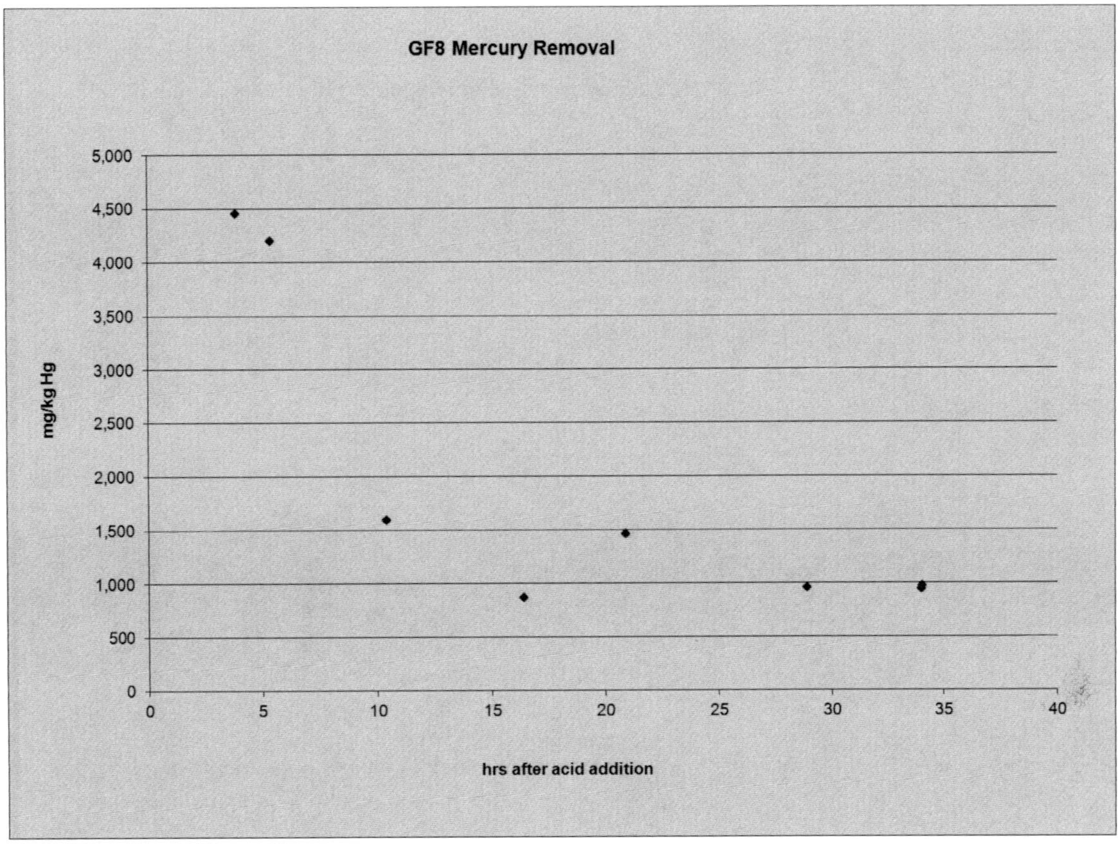

A.2 Rheology Flow Curves

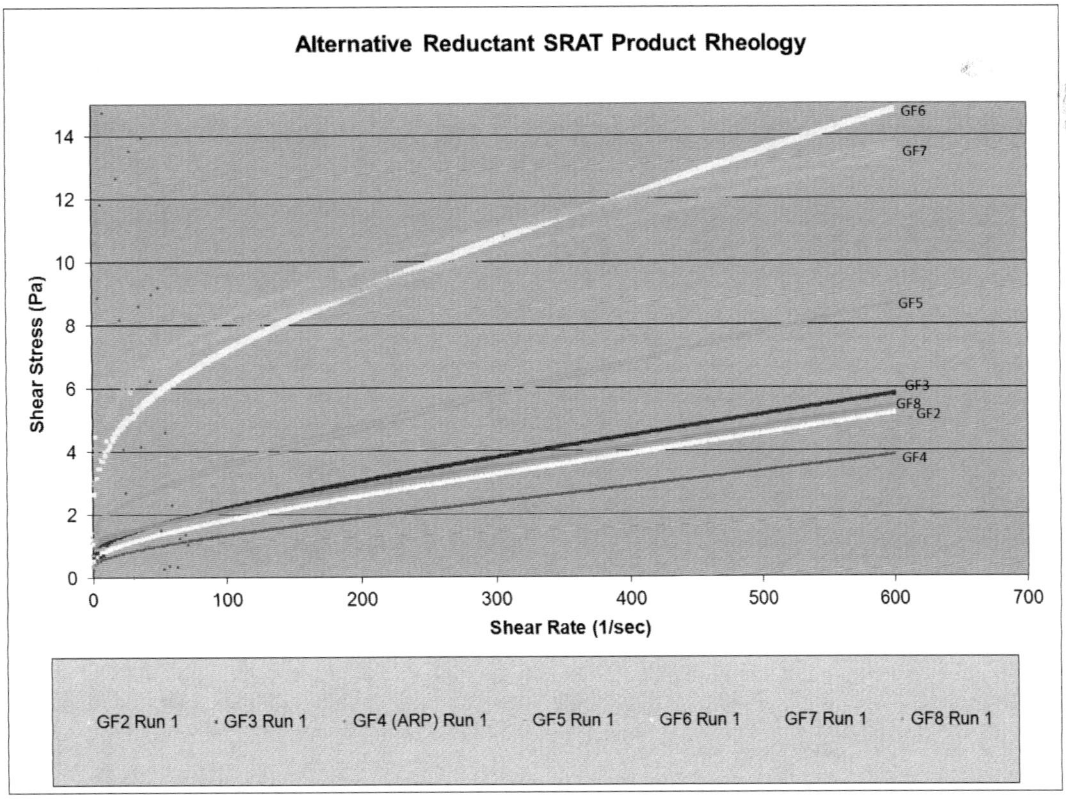

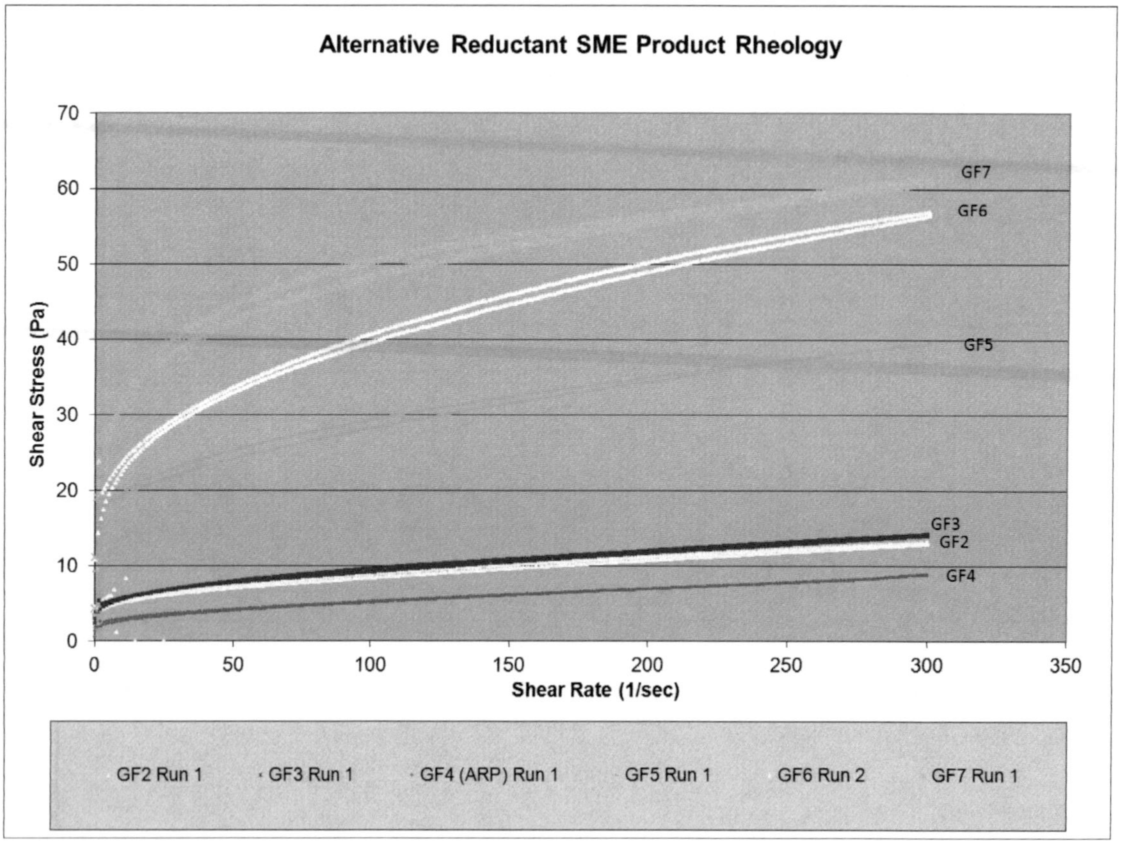
Alternative Reductant SME Product Rheology
Shear Stress (Pa)
Shear Rate (1/sec)
70
60
50
40
30
20
10
0
0
50
100
150
200
250
300
350
GF7
GF6
GF5
GF3
GF2
GF4
GF2 Run 1
GF3 Run 1
GF4 (ARP) Run 1
GF5 Run 1
GF6 Run 2
GF7 Run 1

A.3 Additional GC Data

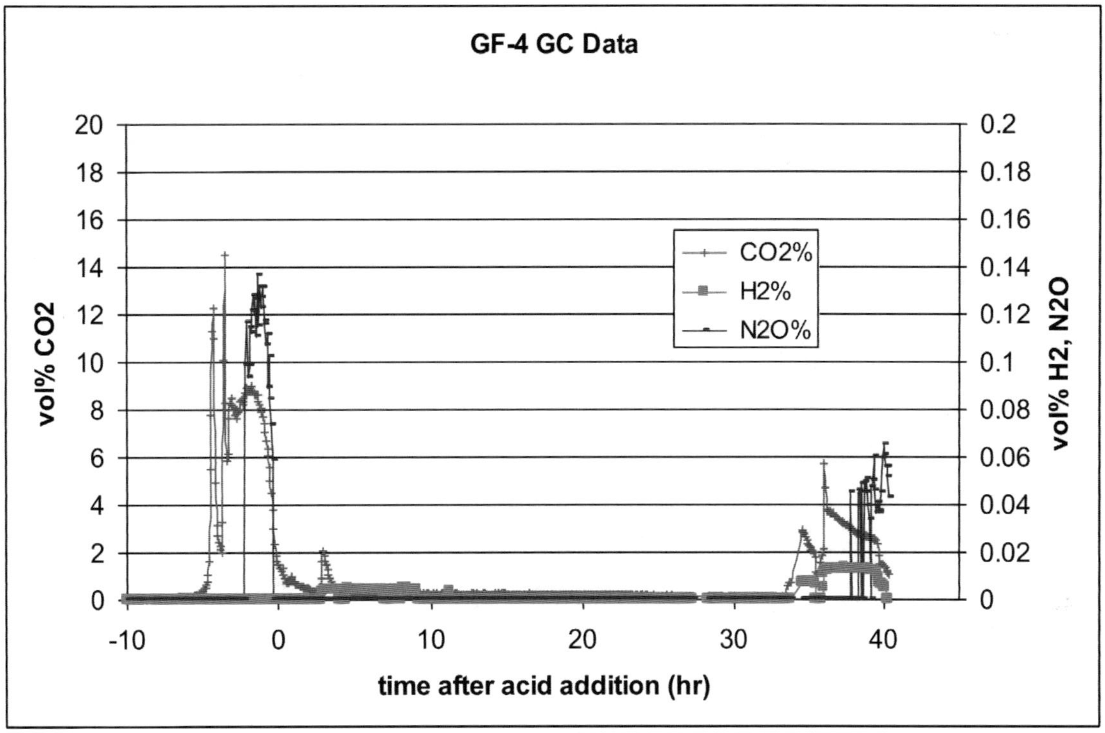

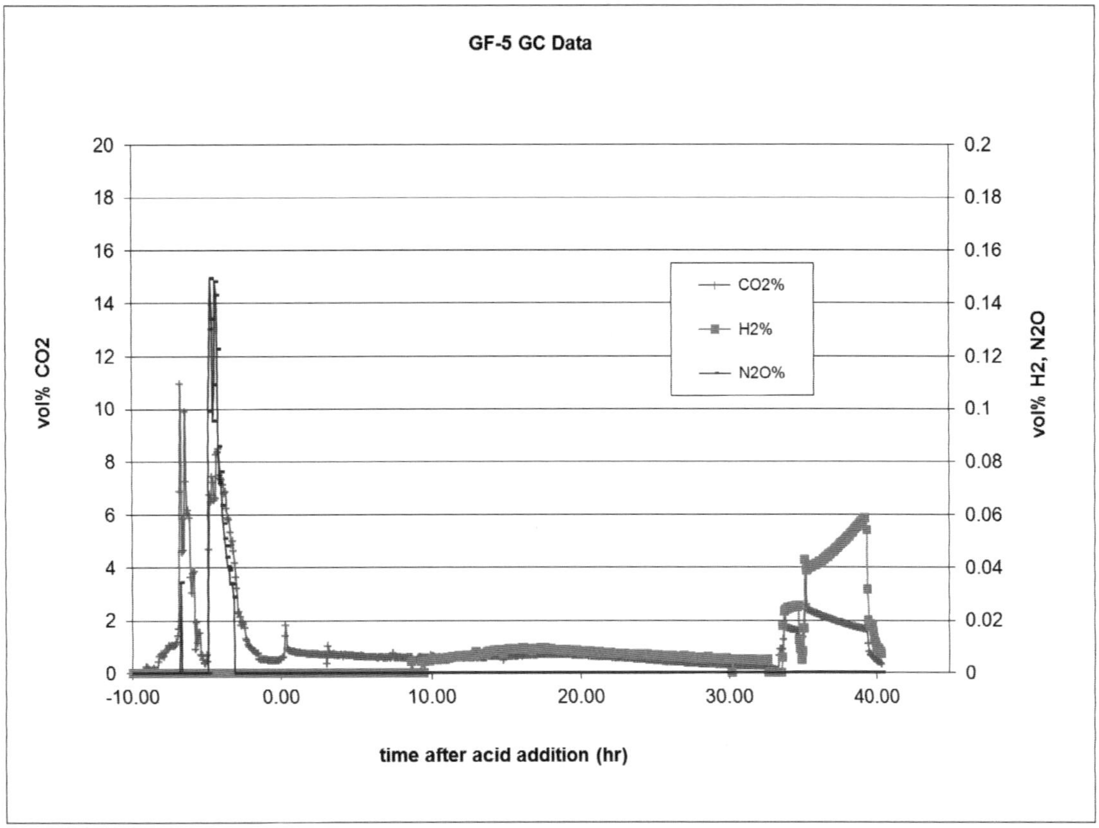

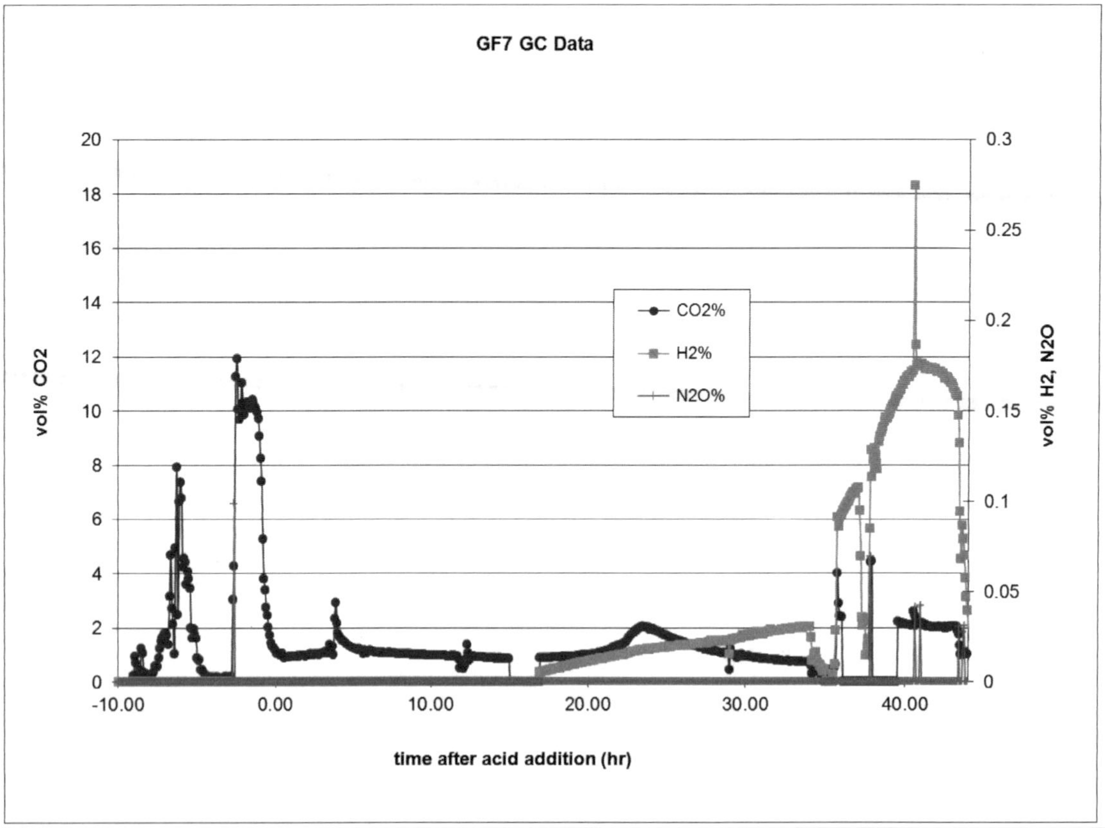
GF7 GC Data
vol% CO2
vol% H2, N2O
CO2%
H2%
N2O%
time after acid addition (hr)
20
18
16
14
12
10
8
6
4
2
0
0.3
0.25
0.2
0.15
0.1
0.05
0
-10.00
0.00
10.00
20.00
30.00
40.00

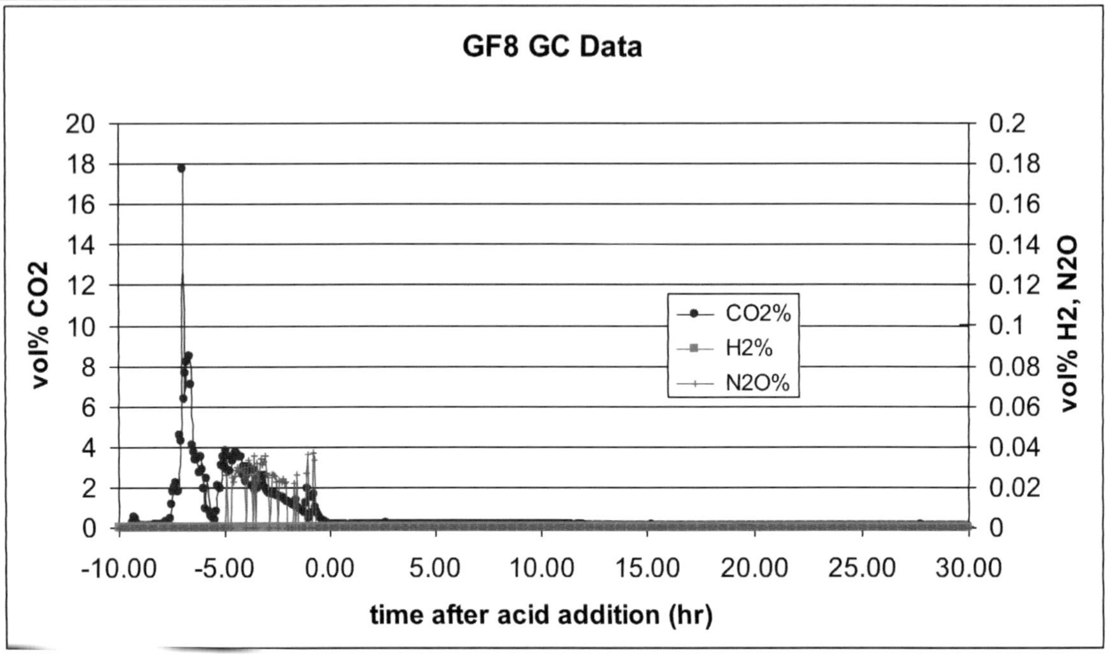
GF8 GC Data
vol% CO2
vol% H2, N2O
CO2%
H2%
N2O%
time after acid addition (hr)
20
18
16
14
12
10
8
6
4
2
0
0.2
0.18
0.16
0.14
0.12
0.1
0.08
0.06
0.04
0.02
0
-10.00
-5.00
0.00
5.00
10.00
15.00
20.00
25.00
30.00

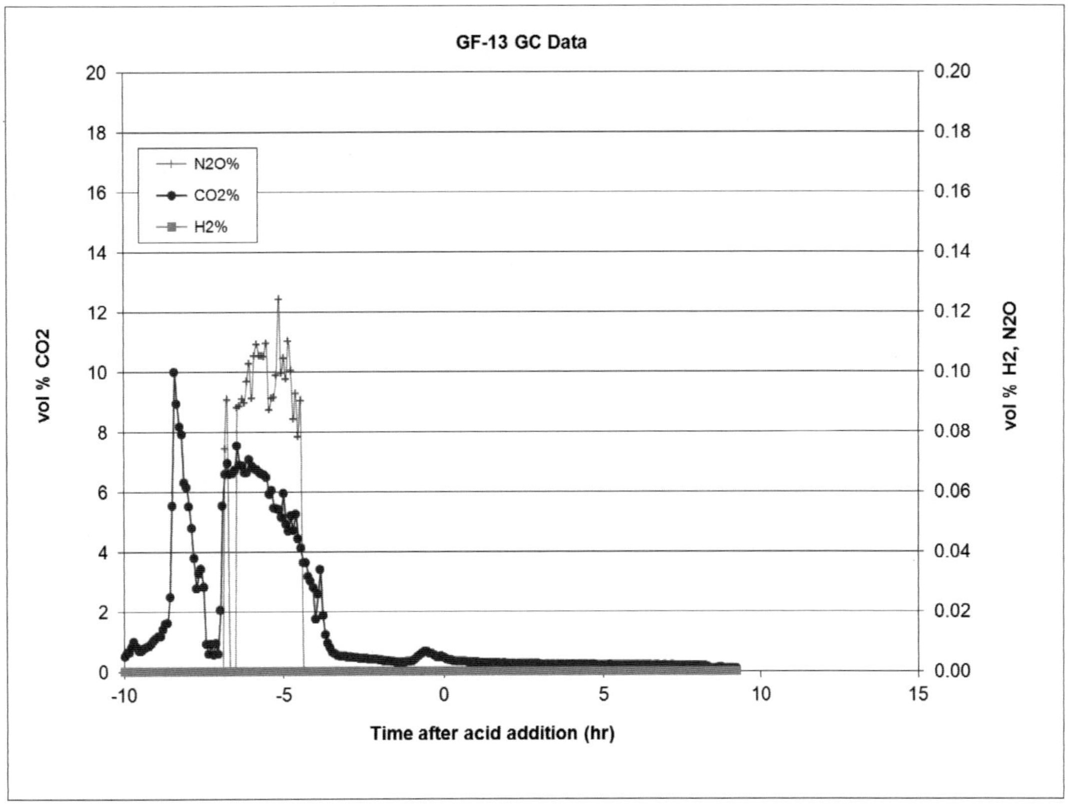
GF-13 GC Data
N2O%
CO2%
H2%
vol % CO2
vol % H2, N2O
Time after acid addition (hr)
20
18
16
14
12
10
8
6
4
2
0
0.20
0.18
0.16
0.14
0.12
0.10
0.08
0.06
0.04
0.02
0.00
-10
-5
0
5
10
15

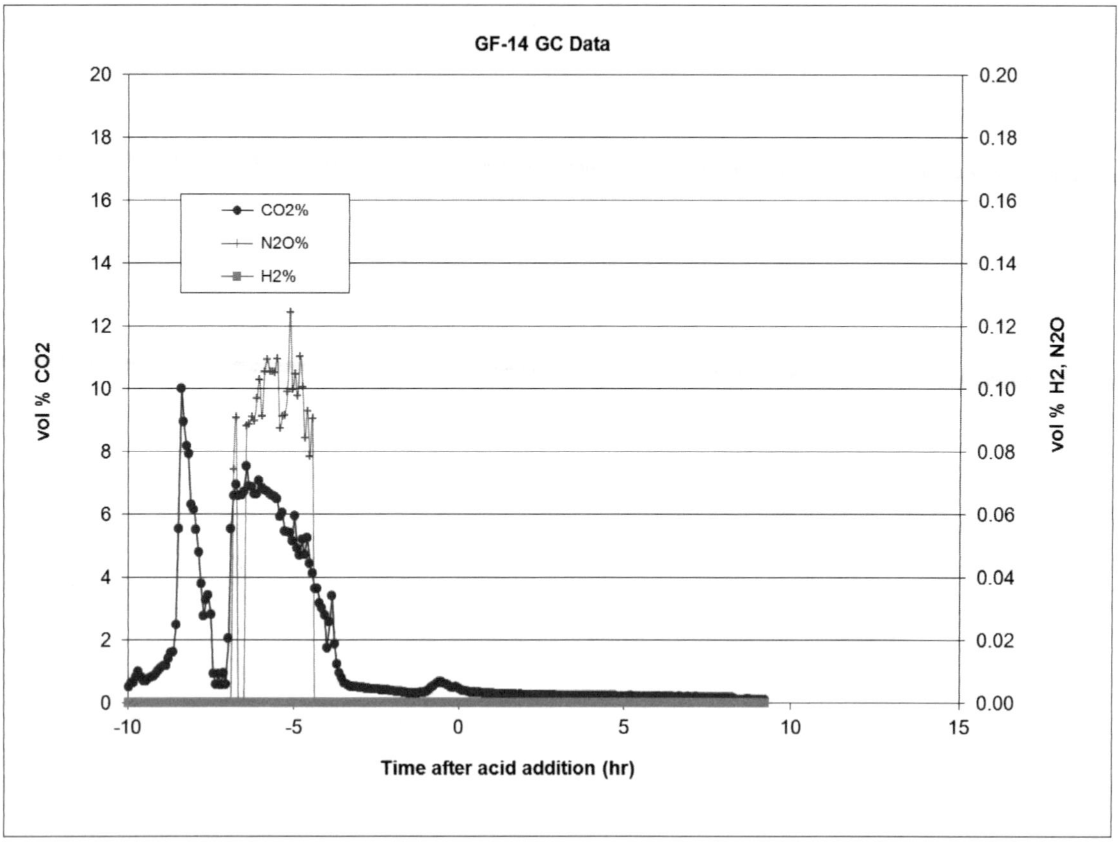
GF-14 GC Data
CO2%
N2O%
H2%
vol % CO2
vol % H2, N2O
Time after acid addition (hr)
20
18
16
14
12
10
8
6
4
2
0
0.20
0.18
0.16
0.14
0.12
0.10
0.08
0.06
0.04
0.02
0.00
-10
-5
0
5
10
15

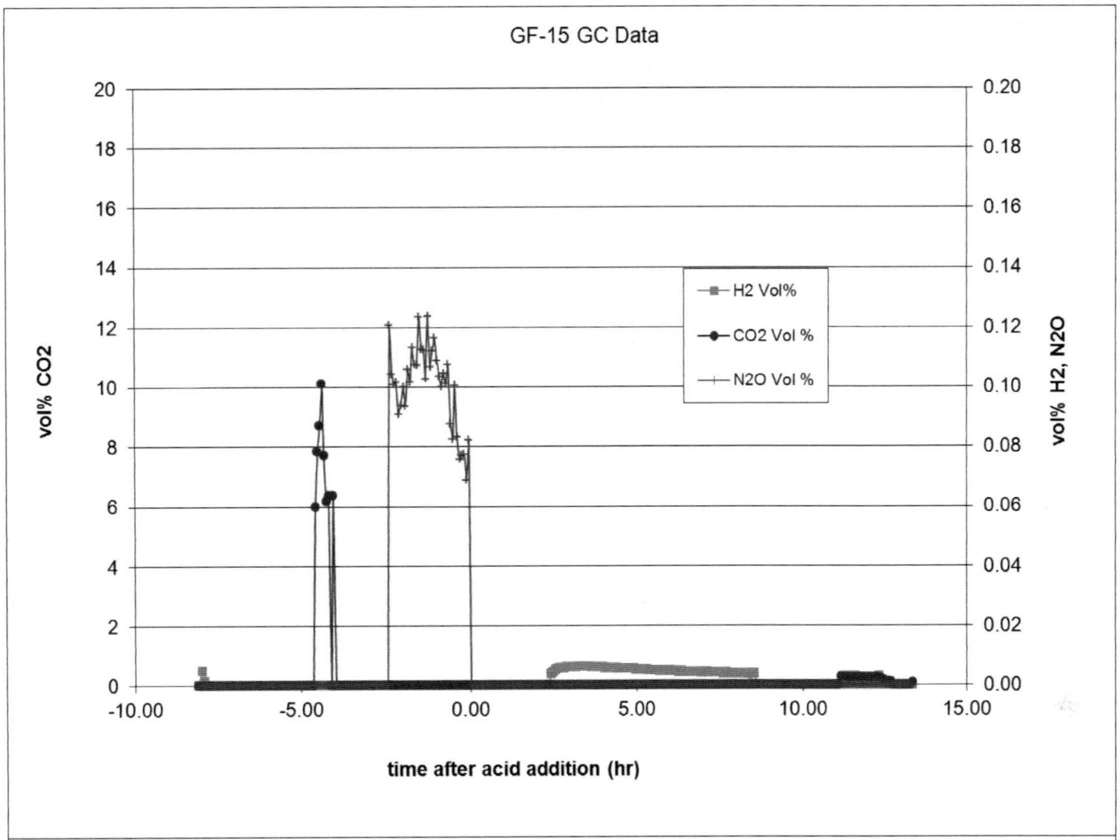
GF-15 GC Data
vol% CO2
vol% H2, N2O
H2 Vol%
CO2 Vol %
N2O Vol %
20
18
16
14
12
10
8
6
4
2
0
0.20
0.18
0.16
0.14
0.12
0.10
0.08
0.06
0.04
0.02
0.00
-10.00
-5.00
0.00
5.00
10.00
15.00
time after acid addition (hr)

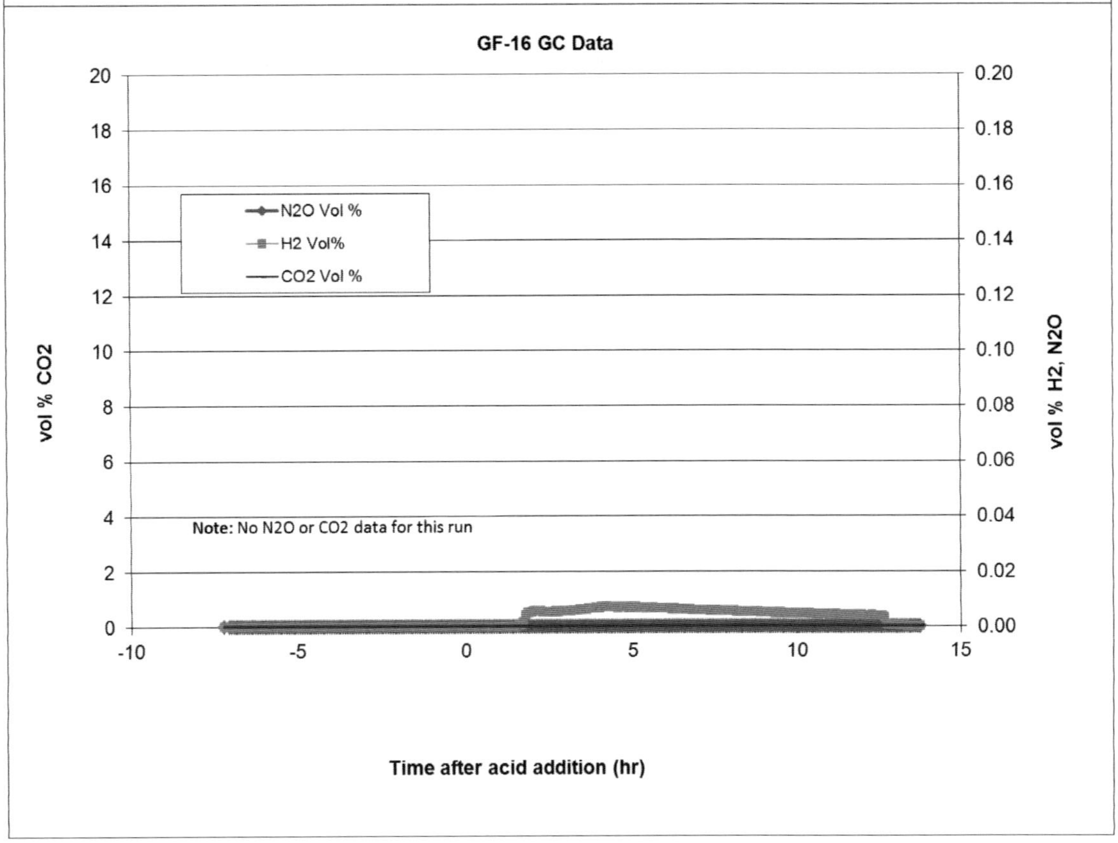
GF-16 GC Data
vol % CO2
vol % H2, N2O
N2O Vol %
H2 Vol%
CO2 Vol %
Note: No N2O or CO2 data for this run
20
18
16
14
12
10
8
6
4
2
0
0.20
0.18
0.16
0.14
0.12
0.10
0.08
0.06
0.04
0.02
0.00
-10
-5
0
5
10
15
Time after acid addition (hr)

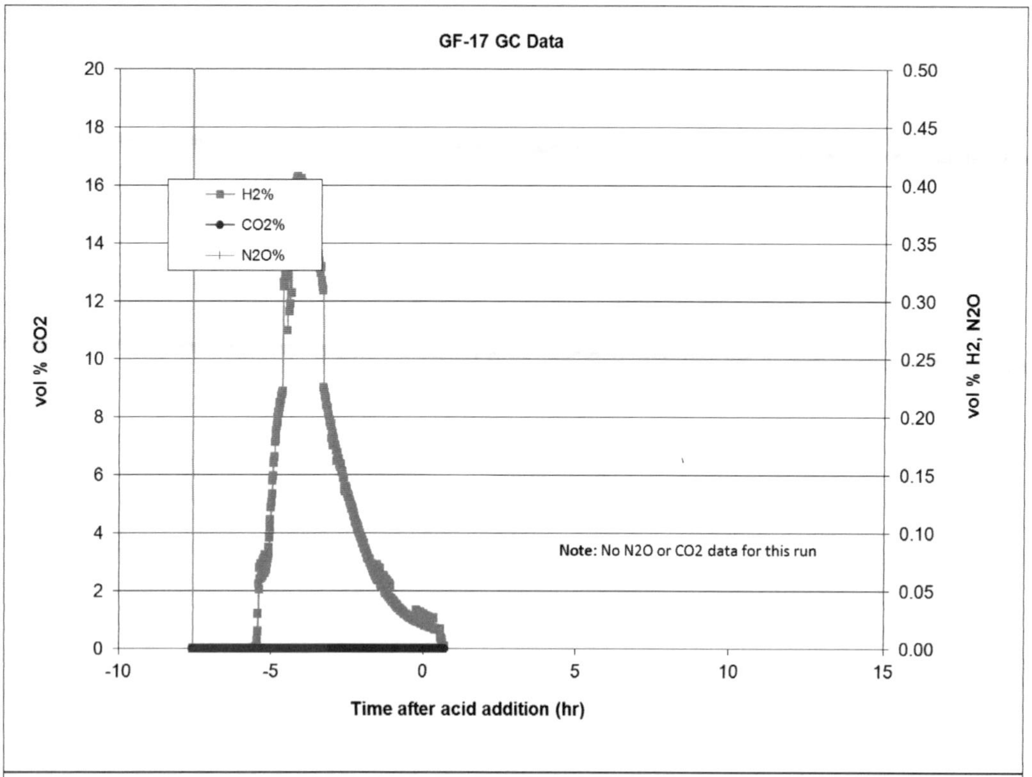
GF-17 GC Data
H2%
CO2%
N2O%
Note: No N2O or CO2 data for this run
vol % CO2
vol % H2, N2O
Time after acid addition (hr)
20
18
16
14
12
10
8
6
4
2
0
0.50
0.45
0.40
0.35
0.30
0.25
0.20
0.15
0.10
0.05
0.00
-10
-5
0
5
10
15

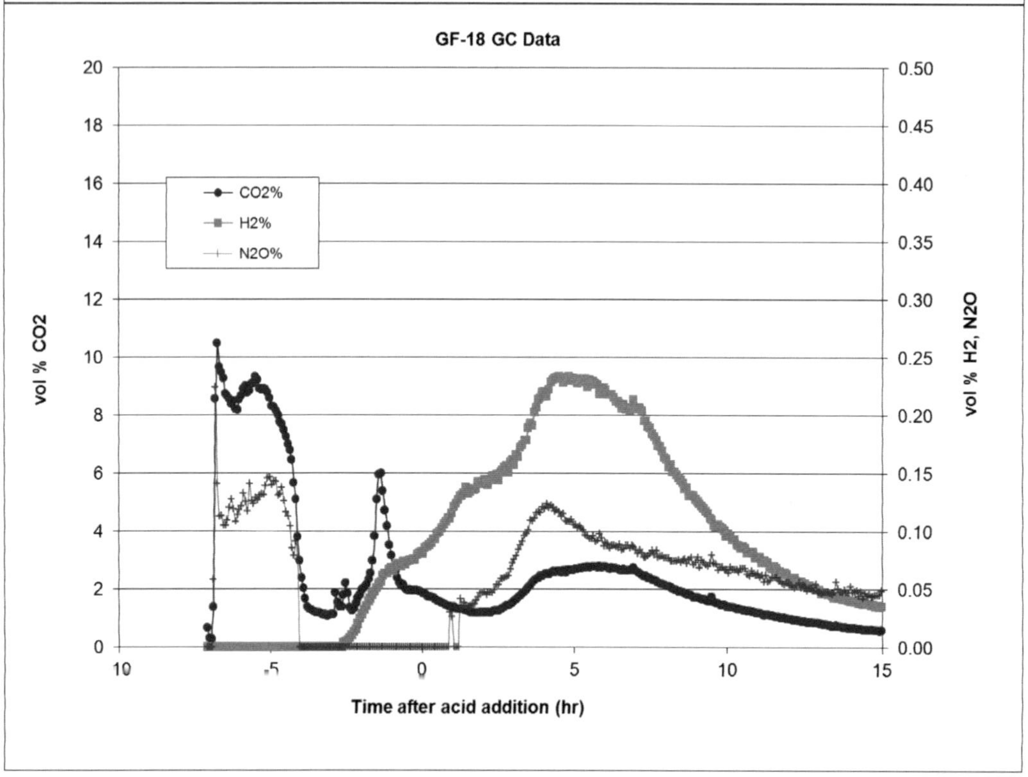
GF-18 GC Data
CO2%
H2%
N2O%
vol % CO2
vol % H2, N2O
Time after acid addition (hr)
20
18
16
14
12
10
8
6
4
2
0
0.50
0.45
0.40
0.35
0.30
0.25
0.20
0.15
0.10
0.05
0.00
10
0
5
10
15

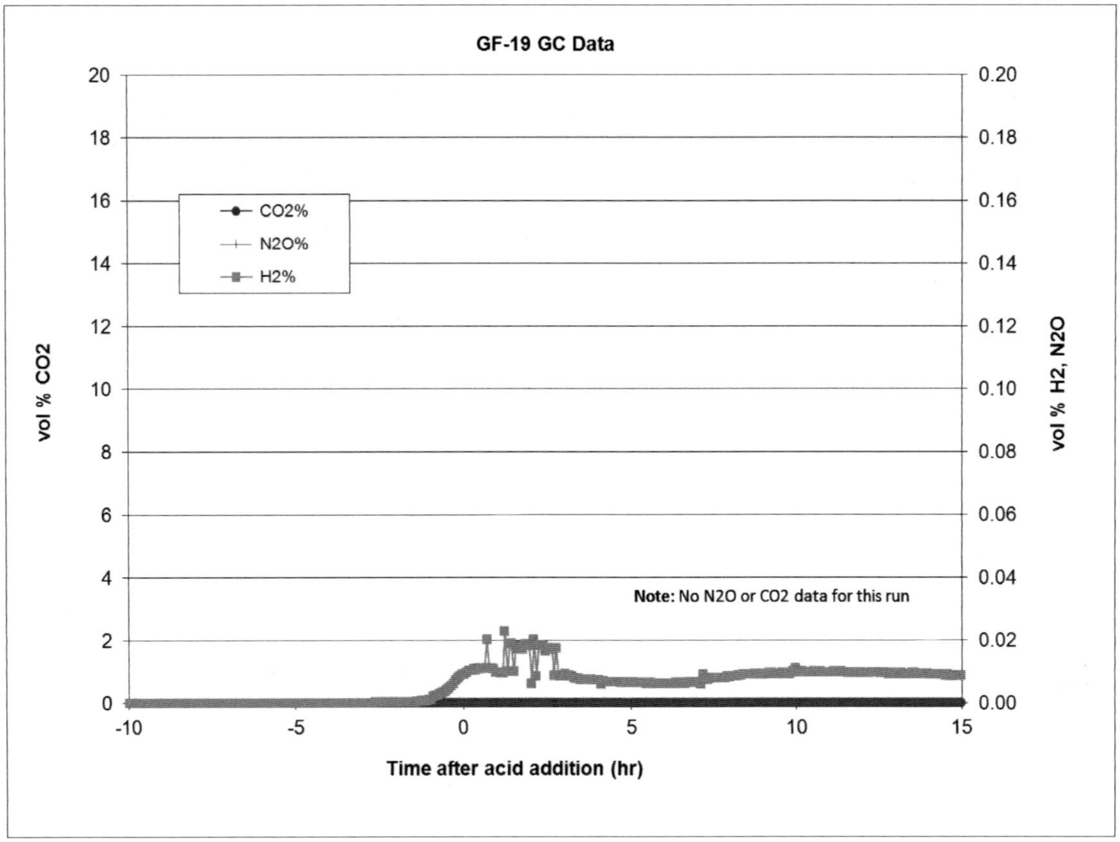
GF-19 GC Data
20
18
16
14
12
10
8
6
4
2
0
0.20
0.18
0.16
0.14
0.12
0.10
0.08
0.06
0.04
0.02
0.00
CO2%
N2O%
H2%
vol % CO2
vol % H2, N2O
Note: No N2O or CO2 data for this run
-10
-5
0
5
10
15
Time after acid addition (hr)

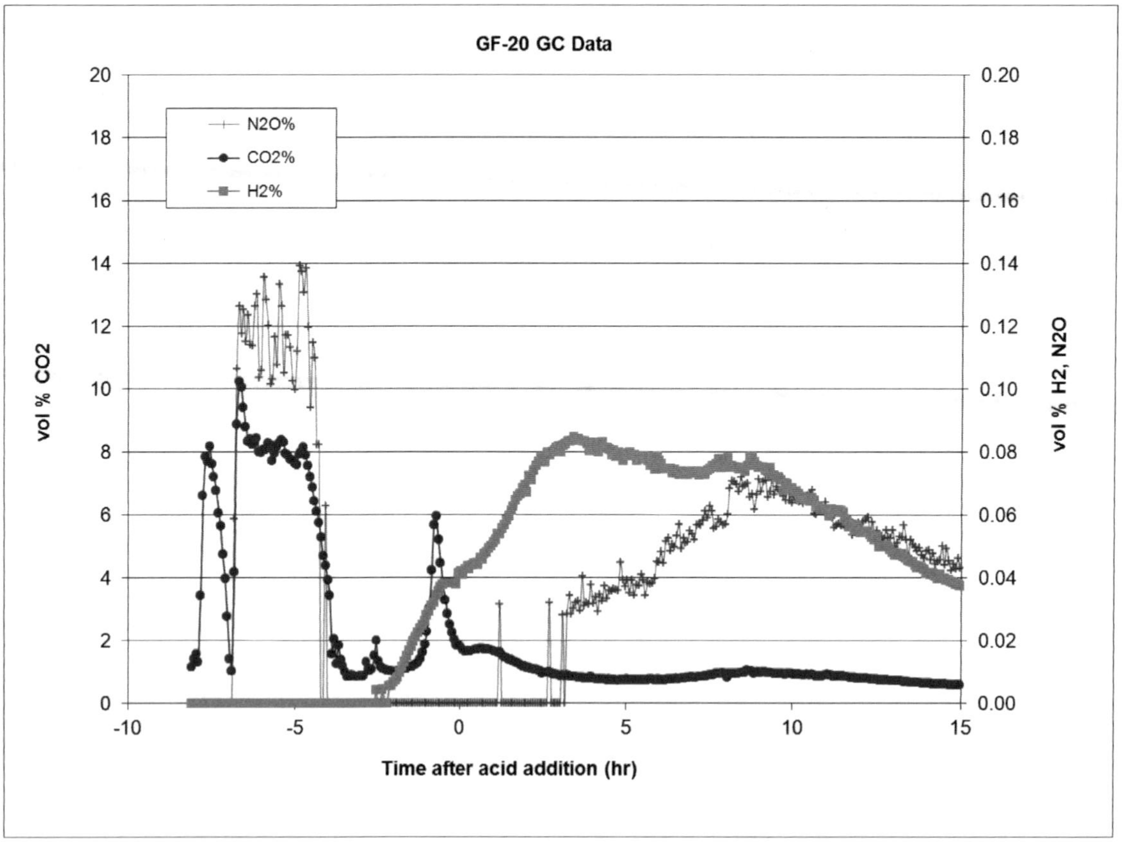
GF-20 GC Data
N2O%
CO2%
H2%
vol % CO2
vol % H2, N2O
Time after acid addition (hr)
20
18
16
14
12
10
8
6
4
2
0
0.20
0.18
0.16
0.14
0.12
0.10
0.08
0.06
0.04
0.02
0.00
-10
-5
0
5
10
15

A.4 Mass Balance Data

GF1 SRAT Carbon Balance

	Input					Calculation						
	Mass, g	TIC, mg/kg	Oxalate, mg/kg	Glycolic Acid, wt %	Formic Acid, wt %	TIC, g	Oxalate, g	Glycolate, g	Formate, g	CO, g	CO2, g	Total C, g
MW						12.0107	88.019	74.03548	45.01744	28.0101	44.0095	12.0107
Number of Carbons						1	2	2	1	1	1	1
SRAT Receipt	3,102.64	1000	815			3.103	2.529	0.000	0.000			3.793
Added Acid	202.20			0.00	90.16			0.000	178.316			47.575
Total In						**3.103**	**2.529**	**0.000**	**178.316**	**0.000**	**0.000**	**51.36769**
	Mass, g	Carbonate, mg/kg	Oxalate, mg/kg	Glycolate, mg/kg	Formate, mg/kg	TIC, g	Oxalate, g	Glycolate, g	Formate, g	CO, g	CO2, g	Total C, g
Offgas											108.96	29.736
SRAT Product	2,481.38		296	0	12250	0.000	0.734	0.000	30.397			8.310
Composite Condensate	1783.60		0	0	322	0.000	0.000	0.000	0.574			0.153
Total Out						**0.000**	**0.734**	**0.000**	**30.971**	**0.000**	**108.960**	**38.20006**

GF3 SRAT Carbon Balance

	Input					Calculation						
	Mass, g	TIC, mg/kg	Oxalate, mg/kg	Glycolic Acid, wt %	Formic Acid, wt %	TIC, g	Oxalate, g	Glycolate, g	Formate, g	CO, g	CO2, g	Total C, g
MW						12.0107	88.019	74.03548	45.01744	28.0101	44.0095	12.0107
Number of Carbons						1	2	2	1	1	1	1
SRAT Receipt	3,127.92	1000	815			3.128	2.549					3.824
Added Acid	278.10			68.25	10.32			184.768	28.072			67.439
Total In	**3406.02**					**3.128**	**2.549**	**184.768**	**28.072**			**71.26264**
	Mass, g	Carbonate, mg/kg	Oxalate, mg/kg	Glycolate, mg/kg	Formate, mg/kg	TIC, g	Oxalate, g	Glycolate, g	Formate, g	CO, g	CO2, g	Total C, g
Offgas											40.24	10.981
SRAT Product	2,422.83		2405	57250	0	0.000	5.827	138.707	0.000			46.595
Composite Condensate	1793.12			62.8	293	0.000	0.000	0.113	0.525			0.177
Total Out	**4215.958**					**0.000**	**5.827**	**138.820**	**0.525**	**0.000**	**40.238**	**57.75306**

GF1 SRAT Nitrogen Balance

	Mass, g	Nitrate, mg/kg	Nitrite, mg/kg	Nitric Acid, wt %	Nitrate, g	Nitrite, g	Ammonium, g	NO, g	N2O, g	Calculated NO2, g	Total N, g
MW					62.0049	46.0055	18.03846	30.0061	44.0128	46.0055	14.0067
Number of Nitrogens					1	1	1	1	2	1	1
SRAT Receipt	3,102.64	6135	16750		19.035	51.969					20.122
Added Acid	202.20			50.03	99.542						22.486
Total In					**118.577**	**51.969**	**0.000**	**0.000**	**0.000**	**0.000**	**42.608**
	Mass, g	Nitrate, mg/kg	Nitrite, mg/kg	Ammonium, mg/L	Nitrate, g	Nitrite, g	Ammonium, g	NO, g	N2O, g	Calculated NO2, g	Total N, g
Offgas								0.000	1.145	60.178	19.051
SRAT Product	2,481.38	20600			51.116	0.000	0.000				11.547
Composite Condensate	1,783.60	156			0.278	0.000					0.063
Scrubber Post SRAT	750.00			7110			5.333				4.141
FAVC Post SRAT	39.40			10600			0.418				0.324
Total Out					**51.395**	**0.000**	**0.418**	**0.000**	**1.145**	**60.178**	**35.126**

GF3 SRAT Nitrogen Balance

	Mass, g	Nitrate, mg/kg	Nitrite, mg/kg	Nitric Acid, wt %	Nitrate, g	Nitrite, g	Ammonium, g	NO, g	N2O, g	Calculated NO2, g	Total N, g
MW					62.0049	46.0055	18.03846	30.0061	44.0128	46.0055	14.0067
Number of Nitrogens					1	1	1	1	2	1	1
SRAT Receipt	3,127.92	6135	16750		19.190	52.393					20.286
Added Acid	224.80			50.03	110.668						24.999
Total In					**129.858**	**52.393**	**0.000**				**45.286**
	Mass, g	Nitrate, mg/kg	Nitrite, mg/kg	Ammonium, mg/L							0
Offgas								1.932	0.000	35.431	11.689
SRAT Product	2,422.83	52650			127.562	0.000	0.000				28.816
Composite Condensate	1,793.12	132			0.237	0.000					0.053
Scrubber Post SRAT	750.00			461			0.346				0.268
FAVC Post SRAT	39.40			883			0.035				0.027
Total Out					**127.799**	**0.000**	**0.035**	**1.932**	**0.000**	**35.431**	**40.827**

(498e) Hydrogen Production From Deliquescence of Ammonia Borane

Chang-Chen Chou[1], Cheng-Hong Liu[1], **Bing-Hung Chen**[1] and Duu-Jong Lee[2], (1)Department of Chemical Engineering, National Cheng Kung University, Tainan, Taiwan, (2)Department of Chemical Engineering, National Taiwan University of Science and Technology, Taipei, Taiwan

Chemical hydride is one of the most potential hydrogen storage materials which can effectively increase hydrogen storage capacity. To meet the new goal of U.S. department of energy (DoE), 4.5wt% and 5.5 wt% of gravimetric hydrogen density capacity by 2010 and 2015, respectively, ammonia borane (AB) is an attractive material due to high intrinsic hydrogen density (i.e. 19.6 wt%). Such a high intrinsic hydrogen density increases the possibility to apply to be the on-board power source of vehicle system, not only for 3C electric product.

In this study, hydrogen evolution from hydrolysis of ammonia borane in limiting water conditions is studied. The solid composite of ammonia borane hydrogen fuel is synthesized with pulverized catalyst through the high energy ball milling process. Various parameters such as the amount of water dosage, temperature, loading of catalysts, on hydrogen production characteristics will be discussed. Kinetic behavior of the hydrolysis reaction in such limiting water supply conditions will also be investigated. Based on experimental outcomes, optimal conditions on hydrogen production from this hydrogen storage system will be proposed.

See more of this Session: Aspects of Hydrogen Storage Materials
See more of this Group/Topical: International Congress on Energy 2011

(343b) Reflections From a Wilson Award Winner

William F. Holcomb, U.S. Government, Montgomery Village, MD

Relections of a chemical engineer working in the nuclear field in the last century and a possible thought on the future.

See more of this Session: Wilson Award Plenary Session on Nuclear Chemical Engineering (Invited Papers)
See more of this Group/Topical: Nuclear Engineering Division

(213d) Isolation and Characterization of a TNT-Degrading Bacterium

Liangdong Zou, Diannan Lu and Zheng Liu, Department of chemical engineering, Tsinghua University, Beijing, China

2,4,6-Trinitrotolene(TNT) is one extensively used explosive for various purposes and is harmful for soil and water. In the present study, four strains named as Thu-A, Thu-B, Thu-C and Thu-Z, were isolated from contaminated sites. Thu-Z had the highest degradation efficiency compare with other strains. It was identified by 16S rDNA assay that Thu-Z was a ***Pantone*** spp. The similarities in both biochemical properties and morphology to the representatives of ***Pantoea agglomerans*** suggested Thu-Z was ***Pantoea agglomerans*** strain. Thu-Z could use TNT as sole nitrogen source while cleavage NO_2 from nitroaromatic ring via direct aromatic ring reduction. Under nitrogen-limited condition, 96.6%N of TNT was used by Thu-Z for growth and with $NaNO_2$ as sole nitrogen source were done as compare. Also trace nitro reduction metabolites were identified. This meant that a dual pathway was found in Thu-Z, and direct aromatic ring reduction was predominant. Addition of other nitrogen source ($(NH_4)_2SO_4$) stimulated the growth of Thu-Z and accelerated TNT degradation.

See more of this Session: Contaminant Transport and Site Remediation
See more of this Group/Topical: Environmental Division

(498b) A New Metal-Decorated Carbon Catalyst for Dehydrogenation of Complex Hydride Hydrogen Storage Materials

Sean S.-Y. Lin[1], Jun Yang[2] and Harold H. Kung[1], (1)Chemical and Biological Engineering, Northwestern University, Evanston, IL, (2)Fuel Cell and Hybrid Electric Vehicle Department, Ford Motor Company, Dearborn, MI

In recent years, many new solid-state hydrides with a high hydrogen capacity that meet DOE requirement (> 7.5 wt.% H_2) have been synthesized as promising on-board hydrogen storage materials.

In these hydrides, however, only a small fraction of hydrogen can be released at a desired on-board operating temperature (<85°C) due to a slow dehydrogenation kinetics of the hydrides. It has been discovered that the kinetics can be improved by adding transition metal catalysts to form metal-hydride intermediates that facilitate hydrogen transportation and desorption. Nevertheless, identification of such catalyst for various hydrides is a challenging task because the catalytic effect is component-specific. Here we report a new form of catalyst for dehydrogenation of complex hydride, using a well-studied $NaAlH_4$ as a test case. Dehydrogenation of $NaAlH_4$ can be greatly facilitated by activated carbon catalysts. The catalytic function can be further enhanced by decorating the carbon with Co, Ni, or Cu nanoparticles. The decomposition temperature was lowered by as much as 100°C using a 3 wt.% Co or Ni-decorated activated carbon, comparable to a Ti-based catalyst, which were the most effective among the metals tested. The catalytic effect is likely due to a combination of hydrogen spillover effect, high contact area between carbon and the hydride, and confinement of the hydride as nano-sized domains in the pores of the carbon matrix. The catalysts were also effective in facilitating rehydrogenation of $NaAlH_4$ under moderate pressure (75.8 bar H_2) and low temperature (120°C), when no rehydrogenation would occur without the catalyst. The fact that this new catalyst system is not specific to any hydride offers many potential applications.

See more of this Session: Aspects of Hydrogen Storage Materials
See more of this Group/Topical: International Congress on Energy 2011

MODELING EXTRACTION OF URANYL NITRATE, NITRIC ACID AND WATER BY TRIBUTYL PHOSPHATE

Stepan Hlushak, Chemical and Biomolecular Engineering, Vanderbilt University, USA, Laboratoire PECSA, Université Pierre et Marie Curie, France, Institute for Condensed Matter Physics, Ukraine

Jean-Pierre Simonin, Laboratoire PECSA, Université Pierre et Marie Curie, France

Olivier Bernard, Laboratoire PECSA, Université Pierre et Marie Curie, France

Philippe Moisy, CEA Marcoule, France

Abstract

The model developed by Naganawa and Tachimori (H. Naganawa, S. Tachimori, Bull. Chem. Soc. Jap. 70, 809, (1997)) was extended to describe extraction equilibria between aqueous phases containing uranyl nitrate and nitric acid contacted with tributyl phosphate (TBP) in dodecane. The introduction of 6 complexes allows one to describe extraction data for the three components (water, nitric acid and uranyl nitrate) in the case of an organic phase composed of 30% in volume of TBP at 25°C. The thermodynamic activities of the components in the aqueous phase were obtained using the extended Pitzer expression. This method gives a good representation of the osmotic coefficient for pure binary aqueous solutions of uranyl nitrate and nitric acid up to 4 mol L^{-1} and 10 mol L^{-1}, respectively. The stoichiometries of the complexes in organic phase were obtained one by one, by an optimisation procedure. The present treatment suggests the formation of complexes containing 2 and 3 TBPs for 1 uranyl nitrate.

Introduction

Nuclear reprocessing uses chemical procedures to separate and recover the useful components, like the remaining uranium and the newly-created plutonium, from the fission products and other radioactive wastes in spent nuclear fuel obtained from nuclear reactors.

Reprocessing serves multiple purposes. Originally reprocessing was used solely to extract plutonium for producing nuclear weapons, but nowadays it is used mainly for reduction of the volume and radiotoxicity of the waste, allowing separate management of nuclear waste components.

The reprocessed uranium, which constitutes about 95% of the mass of the spent fuel material, can be reused as fuel, but it is mainly extracted to decrease the amount of the waste, since the former use is only economic when uranium prices are high. The other 5% of the waste are composed of 4% of fission products and neutron capture products. They are extracted to decrease the radiotoxicity of the waste.

Among the various actinide separation methods developed, solvent extraction has several advantages for use in industrial nuclear reprocessing, including the ability for continuous operation, high throughput and solvent recycling [1]. It constitutes the essence of the PUREX (Plutonium - URanium EXtraction) process, which is the standard aqueous nuclear reprocessing method for the recovery of uranium and plutonium from spent nuclear fuel.

The PUREX process may be roughly described as the extraction of plutonium and uranium into an organic solvent composed of 30% tributyl phosphate (TBP) in odourless kerosene or dodecane from aqueous solution containing nitric acid. The fission products remain in an aqueous phase. Once separated from the fission products, further processing allows separation of the heavier plutonium from the uranium. The PUREX extraction process is based on a solvent extraction process in which complexes are formed between the TBP and the extracted components.

Therefore, a model for the extraction of the aqueous phase components is of interest for the nuclear reprocessing industry. It should be noticed that the currently available models capable of describing the extraction of the water, the acid and the metal seem scarce [2,3].

The aim of this work was to acquire new experimental data and to develop a quantitative model capable of describing the extraction of uranyl nitrate, nitric acid and water simultaneously. The formulation of the model was motivated by a previous successful approach in the case of nitric acid extraction by Naganawa and Tachimori [4]. These authors described the extraction of nitric acid and water in a wide range of TBP concentrations, by assuming the presence of 8 different complexes forming a thermodynamically ideal solution. The equilibrium formation constants of the complexes were found to be independent of the TBP concentration. Thus, it appeared appealing to examine whether this chemical model could be successful when extended to the extraction of uranyl nitrate.

Methodology

Description of the model

We consider the extraction of water, nitric acid and uranyl nitrate by an organic phase composed of TBP diluted in dodecane. Nitric acid and uranyl nitrate are assumed to be present in organic phase only in the form of complexes with TBP. Dodecane and TBP are regarded as totally insoluble in water.

The final solution of the formed complexes is assumed to be ideal. Nonideality in organic phase is assumed to be solely due to association (complex formation) of species with TBP. The complexes formed are denoted by $(TBP)_t{\bullet}(UO_2(HNO_3)_2)_m{\bullet}(HNO_3)_a{\bullet}(H2O)_h$, where *t, m, a* and *h* are the stoichiometric numbers of TBP, uranyl nitrate, nitric acid and water in the complex, respectively. To simplify the notations we will write the complexes in the following form

$$T_tU_mA_aW_h$$

with T, U, A and W standing for TBP, uranyl nitrate, nitric acid and water, respectively.

The composition of the organic phase may be described in terms of true mole fractions of components and complexes. The latter are denoted by $\zeta_{t,m,a,h}$ and are defined as

$$\zeta_{t,m,a,h} = \frac{C_{t,m,a,h}}{C_0 + \sum C_{t,m,a,h}} \tag{1}$$

where $C_{t,m,a,h}$ is the concentration of the $T_tU_mA_aW_h$ complex in organic phase. The sum in the denominator of Eq 1 is taken over all the complexes present in the organic phase, including the concentration of free (unbonded) TBP, $C_{T,free} = C_{1,0,0,0}$, and that of free water, $C_{W,free} = C_{1,0,0,0}$, plus the concentration of dodecane, C_0.

The central law used in the analysis is the mass action law applied to the equilibrium

$$m\,U + a\,A + h\,W + t\,T \leftrightarrow T_tU_mA_aW_h \tag{2}$$

in which T is in organic phase. One has that

$$K_{t,m,a,h} = \frac{\zeta_{t,m,a,h}}{(\zeta_T)^t (a_{U,aq.})^m (a_{A,aq.})^a (a_{W,aq.})^h} \quad (3)$$

in which $K_{t,m,a,h}$ is the equilibrium constant for the formation of the complex $T_tU_mA_aW_h$ (Eq 2) and $a_{X,aq}$ is the activity of species X in aqueous phase. Because the solution of complexes in organic phase is assumed to be ideal, the activities of the complexes are equal to their true mole fractions, $a_{t,m,a,h} = \zeta_{t,m,a,h}$..

Application of the model relies on the assumption of a finite number of complexes that are formed in the organic phase. However, the nature of these complexes has not been revealed experimentally. In the absence of such information, it was chosen to determine the minimum set of complexes leading to a good representation of experimental data. For this purpose, an optimisation approach was developed in which the stoichiometries of the complexes were found one by one, by successive approaches. At each step, the stoichiometry of the probable next main complex was found by a fit of experimental data using a differential evolution global optimisation algorithm [5]. Then the newly found complex was added to the set and the equilibrium constants for all complexes were adjusted. The procedure was repeated until a satisfying description of experimental data was obtained.

Experimental

The concentrations of species in the two phases were measured by sampling equilibrated solutions. The latter were obtained by mixing volumes of aqueous and organic phases and centrifuging them, at 25°C. Then, the densities of the separated phases were measured (at 25°C) using an Anton-Paar DMA55 tuning-fork density transducer. The concentrations of uranyl nitrate and nitric acid in the phases were determined by spectrophotometric measurement (using a Hitachi U-3000 spectrophotometer) and acid-base titration, respectively. The concentration of water in organic phase was measured by coulometric Karl Fischer titration with a Metrohm KF 737 titrator.

Results And Discussion

The first part of the work was devoted to the description of extraction of pure nitric acid from the aqueous phase, in the absence of uranyl nitrate. We used the data of Naganawa and Tachimori [4] and refitted them with our model. The concentration of the dodecane was found from the known density of organic phase and concentrations of other components. As in previous work [4], 8 complexes were assumed to form in the organic phase: TW, T_2W_2, T_3W_6, AW, TA, TAW, T_2AW and T_3AW_4. Excellent agreement with experimental results was obtained.

Now, keeping these complexes and their equilibrium formation constants, we looked for the complexes involving uranyl nitrate. Six additional complexes were identified one after another for 30% TBP in dodecane, in the following order: T_2U, T_2UW, T_2UA, T_3UA, T_3UW and T_3U. They are therefore of the form T_nU and T_nUX, with n= 2 or 3 and X= A or W. In the calculations, the activity coefficients of nitric acid and uranyl nitrate in the aqueous ternary mixtures were estimated using the extended Pitzer expression [6]. These quantities can be obtained from the results for the binary solutions without additional parameter. The result for the osmotic coefficient of the binary solutions is shown in Figure 1.

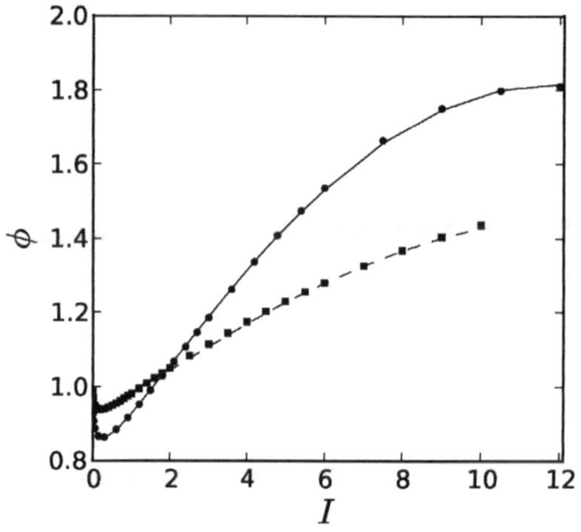

Fig 1. Description of osmotic coefficient for binary aqueous solutions of nitric acid (dash-dotted line) and uranyl nitrate (solid line); symbols indicate experimental values.

The final results for the extraction of uranyl nitrate, nitric acid and water are plotted in Figures 2, 3 and 4, respectively. They show a good representation of data for the three species.

The results exhibit rapid extraction of uranyl nitrate as a function of its concentration in aqueous phase, which seems to result predominantly from the formation of the T_2U complex. This is particularly clear at high aqueous uranyl nitrate concentration, where the concentration of extracted uranyl is independent of aqueous nitric acid activity and approaches 0.5 mol L^{-1}, which is half of the concentration of TBP. Sharp increase of uranyl nitrate extraction is accompanied by strong back extraction of nitric acid and water to the aqueous phase. The latter phenomenon may be explained by a competition between uranyl nitrate, nitric acid and water for association with TBP molecules.

As more uranyl nitrate is added to the aqueous phase, the weak complexes in the organic phase, composed of TBP with nitric acid and water, are superseded by stronger ones with uranyl nitrate. The nitric acid and the water released by the dissociation of weak complexes are then back extracted to aqueous phase. Nonetheless, small amounts of water and nitric acid remain in the organic phase in the form of T_nUW and T_nUA complexes with n= 2 and 3.

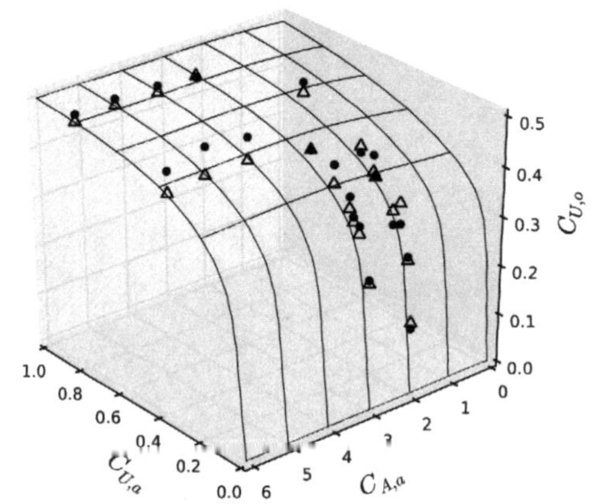

Fig 2. Concentration of uranyl nitrate in organic phase as a function of nitric acid and uranyl nitrate concentrations in aqueous phase. Full circles: experimental results; empty triangles: results from the model.

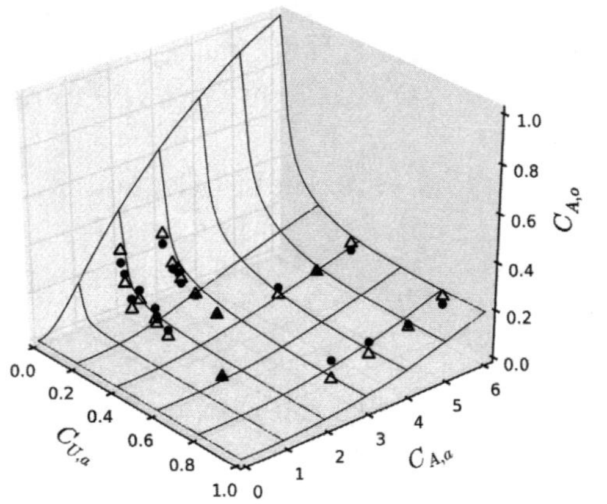

Fig 3. Similar to Figure 1, but for the concentration of nitric acid in organic phase.

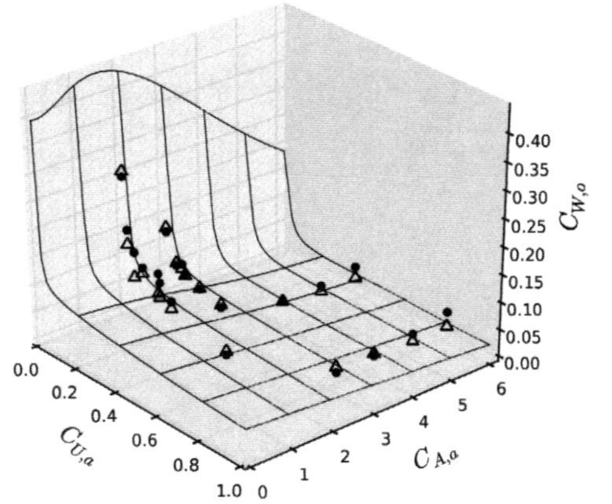

Fig 4. Similar to Figure 1, but for the concentration of water in organic phase.

Conclusions

The ability of a simple thermodynamic model to describe the extraction equilibria of uranyl nitrate, nitric acid and water by TBP in dodecane was examined. In the case of an organic phase composed of 30% TBP, the treatment leads to a good representation of experimental data for various compositions of aqueous phase.

The case of different TBP concentrations however remains to be studied. This will be considered in future work.

The present work suggests the formation of different complexes involving two and three TBP molecules per uranyl nitrate molecule. The complexes with three TBPs are especially relevant at low uranyl nitrate concentration. Some of the complexes of uranyl also include nitric acid and water molecules and are the main complexes responsible for the extraction of water and nitric acid to the organic phase at high uranyl nitrate concentrations.

References

1. Nash, K.L., Madic, C., Mathur, J.N. and Lacquement, J. (2006) "Actinide Separation Science and Technology". *Chemistry of the actinide and transactinide elements*, Springer, vol. 4, pp. 2622-2798.

2. Mokili, B. & Moitrenaud, C. (1996) "Modelling of the Extraction of Neodymium and Praseodymium Nitrates from Aqueous Solutions Containing a Salting-Out Agent or Nitric Acid by Tri-N-butylphosphate." *Solvent Extraction and Ion Exchange* 14(4), pp.617-634.

3. Mokili, B. & Poitrenaud, C. (1997) "Medium Effect on the Separation Factor in Liquid-Liquid Extraction. Application to the Separation of Trivalent Lanthanide Nitrates by Tri-n-Butylphosphate". *Solvent Extraction and Ion Exchange* 15(3), pp.455-481.

4. Naganawa, H. & Tachimori, S. (1997) "Complex Formation between Tributyl Phosphate and Nitric Acid and the Hydration of the Complexes in Dodecane". *Bulletin of the Chemical Society of Japan* 70(4), pp.809-819.

5. Price, K.V., Storn, R.M. and Lampinen, J.A. (2005) "Differential Evolution: a Practical Approach to Global Optimisation". *Springer Verlag.*

6. Goldberg, R.N., Manley, J.L. & Nuttall, R.L. (1985) "GAMPHI - a Database of Activity and Osmotic Coefficients for aqueous electrolyte solutions", *NBS*, Gaithersburg.

(563c) Structure of Rigid Hard-Ring Fluid

Mariam Nouri Dariani and Marc Robert, Chemical and Biomolecular Engineering, Rice University, Houston, TX

The structure of hard ring fluids is studied in two and three dimensions, using Monte Carlo computer simulations in the canonical ensemble. These fluids are composed of molecules consisting of hard-spheres linked together to form planar rigid rings. For rings of various sizes and for a wide range of densities, results are reported for the pair distribution function of the ring centers and for the pair distribution of the ring orientations. In a dense monolayer, where the rings are completely confined to a plane, a "shoulder", precursor of the freezing transition, is observed in the second peak of the pair distribution function of the ring centers, as previously seen in the simple hard-sphere fluid. In two dimensions, where the ring centers are confined to a plane but the rings themselves are free to rotate on all axes, nematic and tetratic ordering is observed at sufficiently high densities. Results are also presented for three dimensions.

See more of this Session: Advances In Numerical Simulations Bridging Chemical and Nuclear Engineering Phenomena or Processes
See more of this Group/Topical: Nuclear Engineering Division

(450d) Automated Systems for Processing of Cyclotron Irradiated Solid Targets

Suzanne E. Lapi, Evelyn Madrid, Thomas Voller, Paul Carey, Maiko Kume, Deborah Sultan and Michael Welch, Radiology, Washington University, St. Louis, MO

Development of agents for Positron Emission Tomography (PET) is a rapidly growing field. The isotopes ^{18}F and ^{11}C are the most commonly used for labeling compounds of interest, but the short half-lives of these isotopes (120 minutes and 20 minutes respectively) often limits the studies which can be conducted to ones which occur on a rapid timescale. The production of longer-lived positron-emitting radionuclides ($t_{1/2}$= hours-days) often requires the use of solid materials as targets. The processing of these solid targets presents different challenges than the liquid targets (used for production of ^{18}F and ^{13}N) or gas targets (used for the production of ^{11}C). We have developed automated systems for the production of ^{64}Cu via the $^{64}Ni(p,n)$ reaction, $^{76,77}Br$ via the $^{76,77}Se(p,n)$ reaction and for ^{86}Y via the $^{86}Sr(p,n)$ reaction. These systems are at different stages of development, with the ^{64}Cu processing system being the most advanced.

^{64}Cu ($t_{1/2}$= 12.7 hours) is an important radioisotope as it decays via both positron emission and negatron emission. Thus it has sparked interest as a potential isotope for both imaging and therapy. We have been producing this isotope for over 15 years and are currently shipping out substantial quantities (200-400 mCi) to some 12-15 institutions on a weekly basis for both preclinical studies and clinical trials. This isotope is produced via irradiation of an enriched ^{64}Ni target plated on a gold disk support. After irradiation, the target material is dissolved and the $^{64}Cu/^{64}Ni$ separation is conducted remotely via ion exchange. The system is designed in such a manner as to recover the ^{64}Ni target material for recycling.

$^{76,77}Br$ are two isotopes which are of interest as an imaging therapy pair. Our group has been producing these radiobromine isotopes for imaging and therapy using brominated estrogen and androgen analogues. The target material is doubly enriched $^{63}Cu_2{}^{76,77}Se$. The purpose of the enriched ^{63}Cu is to prevent the long-lived contaminant ^{65}Zn ($t_{1/2}$ = 244 d) from building up in the system. The radiobromine is extracted via a remote distillation apparatus where the target is heated in a quartz tube furnace to distill the radiobromine which subsequently condenses on a cooler section of the tube. The radioactivity is eluted from the tube and the target can be re-irradiated.

^{86}Y is of interest as an imaging analogue for ^{90}Y therapeutics. This is desirable as ^{90}Y is a pure β emitter and thus understanding the biodistribution and pharmacokinetics of new agents labeled with this isotope is difficult. These compounds can be labeled with ^{86}Y and imaging studies can be carried out before the therapeutic experiments with ^{90}Y. ^{86}Y is produced via proton bombardment of ^{86}Sr in the form of ^{86}SrO. The target material is dissolved and separated using the differences in solubility of strontium and yttrium in acidic and basic solutions. This technique is currently in the process of being automated.

See more of this Session: Engineering Applications of Radioisotopes for Industry and Medicine
See more of this Group/Topical: Nuclear Engineering Division

Removal of ^{14}C from Irradiated Graphite for Waste Volume Reduction and Bulk Graphite Recycle

Tara E. Smith, Shilo McCrory, and Mary Lou Dunzik-Gougar, PhD
Idaho State University: 1776 Science Center Dr., Idaho Falls, ID, 83401
smittar4@isu.edu, mccrshil@isu.edu, mldg@isu.edu

INTRODUCTION

Public concerns regarding availability of energy and environmental health are driving the growth of nuclear energy. The U.S. and other countries are developing advanced nuclear systems coined "Generation IV concepts" as established by the Generation IV International Forum, which institutes protocol for individual countries to lead the development of a concept in which they have particular interest [1]. The U.S. Department of Energy commissioned the Idaho National Laboratory to develop the High Temperature Gas-Cooled Reactor (HTGR). The HTGR uses graphite coated fuel particles, graphite reflector and core structural components, and helium coolant [2]. As graphite is bombarded with neutrons, carbon-14 (^{14}C) is produced through three distinct reactions with ^{13}C, ^{14}N, and ^{17}O. In irradiated reactor graphite, the majority of ^{14}C has been found on the surface [3]. This occurrence indicates the most significant of the production reactions is neutron activation of ^{14}N, because ^{14}N is present on the surface and in surface pores as N_2 gas deposited from air. Carbon-14 is relatively long-lived (half life = 5730 years) and has significant mobility in groundwater and atmospheric systems. For these reasons, disposal of large irradiated graphite components from an HTGR would likely be costly. Further, the value of such pure nuclear–grade graphite is expected to increase and recycle may be an option. Thus, a means of removing the majority surface ^{14}C from the bulk graphite ^{12}C is being investigated. Pyrolysis and oxidation in a steam atmosphere have been suggested as ^{14}C decontamination methods [4]. Fachinger et al. at the Julich Research Center demonstrated the concept of thermal treatment of irradiated nuclear graphite in the presence of steam or oxygen [5]. Thermal treatment combines graphite calcination and chemical treatment to separate non-volatile and volatile radionuclides [4]. However, this method needs to be further refined in order to offer a valid solution for graphite waste management. This paper studies the availability and form of the adsorbed oxygen complexes as preliminary work to refine the thermal treatment process in an effort to optimize the removal of ^{14}C.

DESCRIPTION OF EXPERIMENTAL WORK

Work discussed here is part of a larger project with the objective to determine the chemical nature of the ^{14}C in irradiated graphite and to use this information to determine an optimal method for ^{14}C removal.

Two types of graphite, nuclear grade NBG-18 and POCOFoam®, a highly porous graphite foam developed by Oak Ridge National Laboratory, were selected for analysis. Graphite foam was chosen for its significant surface area, which adsorbed more nitrogen than less porous graphites. It was speculated that higher nitrogen concentrations would result in production of more ^{14}C upon irradiation.

During thermal treatment, irradiated graphite samples are heated to a temperature in the range of 600-1400°C in the presence of inert argon gas, which carries any gaseous products. Graphite gasifies after surface oxidation. Typically oxidation kinetics are a three step process: adsorption, surface reaction, followed by desorption. Thus the effectiveness of ^{14}C removal from irradiated graphite during thermal treatment is highly dependent on the availability of oxygen compounds. Adsorption of oxygen complexes occurs primarily on exposure to air during manufacturing and assembly, and/or exposure to coolant impurities. To learn more about the chemical forms and relative quantities of oxygen compounds present on the pre- and post irradiation graphite samples., surface sensitive techniques were employed: Time-of-Flight Secondary Ion Mass Spectrometry (ToF-SIMS) and X-ray Photoelectron Spectroscopy (XPS).

RESULTS

SIMS identified the most abundant oxygen complex on the pre- irradiated graphite as Oxygen-16 (^{16}O). The presence of ^{16}O is indicative of chemisorptions. While other oxygen-compounds were identified, no carbon-oxygen compounds were found. Therefore adsorbed oxygen compounds are likely weakly chemisorbed to the graphite surface. On irradiated samples, SIMS identified primarily carbon-oxygen compounds, indicating the formation of strong chemisorbed bonds. It is likely that irradiation damage to the graphite significantly increases the number of active sites, and thus dramatically increases the number and strength of the adsorbed oxygen species.

The deconvolution of the O 1s XPS spectrum for irradiated POCOFoam® identified three peaks: peak 1 (533.04 eV), indicative of non-carbonyl (ether-type) oxygen atoms in esters; peak 2 (535 eV), oxygen atoms in carboxyl groups; and peak 3 (536.32 eV), which corresponds to adsorbed water and/or oxygen [15]. Similarly, the deconvolution of the O 1s XPS spectrum for irradiated NBG-18 identified three peaks: peak 1 (531.18 eV), indicative of oxygen atoms from C=O in carboxyl and/or carbonyl groups; peak 2 (532.01 eV), carbonyl oxygen atoms in esters, anhydrides, and/or oxygen atoms in hydroxyl groups; and lastly peak 3 (534.38 eV), oxygen atoms in carboxyl groups [6].

El-Genk *et al.* have speculated that the compounds seen in Figure 1 are likely chemical forms of adsorbed oxygen [7]. The functional group fragments identified by XPS fit El-Genk *et al.'s* speculations. Based on the oxygen functional group stabilities, graphite is expected to desorb primarily carbon monoxide (CO).

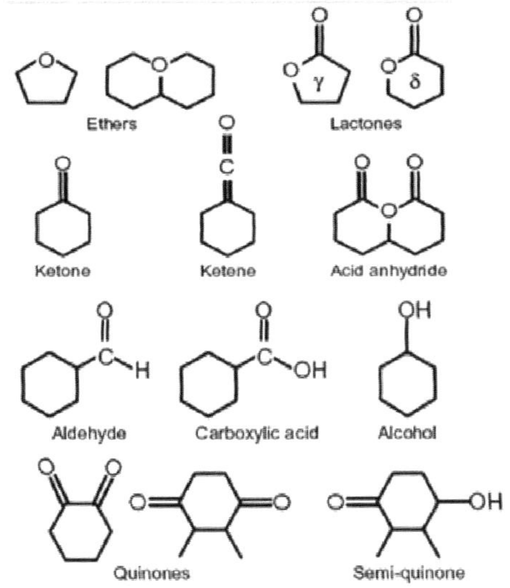

Figure 1. *Plausible formulas for compounds formed on oxygen adsorption on graphite surface as speculated by El-Genk et al. [7]*

The avulsion of CO results after the direct release from an ether or carbonyl; after the decomposition of an ether, ketene, or aldehyde; or after the gasification of surrounding carbon atoms within the graphite microstructure as seen in ethers, lactones, and anhydrides [7]. The desorption of CO can be maximized by heating the graphite up to temperatures above the surface decomposition temperature, which is about 1000°C, so that the ether, carbonyl, ketene, or aldehyde complexes readily breakdown [7,8].

REFERENCES

1. NUCLEAR ENERGY AGENCY. "*Generation IV International Forum.*" Technical Secretariat. 2009
2. Gen-IV International Forum. "*The Very-High-Temperature Reactor (VHTR) is a graphite-moderated, helium-cooled reactor with a thermal neutron spectrum.*" http://www.gen-4.org/Technology/systems/vhtr.htm
3. ELECTRIC POWER RESEARCH INSTITUTE, "*Graphite Decommissioning: Options for Graphite Treatment, Recycling, or Disposal, including a Discussion of Safety-Related Issues,"* EPRI Technical Report 1013091(March 2006)
4. J. FACHINGER, L.V.WERNER, "*Decontamination of Nuclear Graphite,"* 3rd International Topical Meeting on High Temperature Reactor Technology. (October 2006)
5. FACHINGER, J., PODRUHZINA, T., VON LENSA, W., "Decontamination of Nuclear Graphite by Thermal Treatment," Solutions for Graphite Waste, Manchester, UK (March 2007)
6. Zhou, J., Sui, J., Li, P., Chen, D., *"Characterization of Surface Oxygen Complexes on Carbon Nanofibers by TPD, XPS, and FT-IR."* Elsevier, 2006.
7. El-Genk, Mohamed S, Tournier, JMP. *"Development and validation of a model for the chemical kinetics of graphite oxidation."* Journal of Nuclear Materials. January 2011.
8. Watanabe, H., *"Local Etching Effect of Fin Gold Particles Deposited on Graphite."* Japanese Journal of Applied Physics, 1993.

(213e) Chemical Release During Dredging of Contaminated Sediment

R. Ravikrishna, Prathamesh Ekawde and Kanithi Mahesh, Chemical Engineering, Indian Institute of Technology-Madras, Chennai, India

One of the options forthe remediation of contaminated sediments in rivers, lakes or coastal bay areasis dredging and subsequent relocation of the dredged material for storageand/or disposal. Dredging induces resuspension of bedsediment, the extent of which depends on the type of dredging as well as theefficiency of execution. One of the concerns from a risk assessment perspectiveis the release of chemicals into water and possibly to air as a result of desorption of the hazardous chemicals from the contaminatedsolids, which are resuspended [1]. From a riskassessment perspective, it is necessary to evaluate the chemical release duringdredging since this affects water quality and potentially the air quality aswell.

The primary point ofconcern is the rate of chemical release from the resuspendedsolids during dredging by desorption to the waterlayer, where it can become 'bioavailable' orevaporate. Traditionally, the prediction of the magnitude of chemical releaseis based on mass balance measurements of suspended solids concentrations andthe bed sediment chemical concentration assuming .uniform chemical loading inthe undisturbed bed sediment with respect to particle sizes. Previous experiments in laboratorysimulations suggested that the chemical concentration associated with thesolids may be particle size dependent. Since at different times, the suspendedsolids concentration and particle size distribution vary, a model that predictsthe aqueous phase concentration must incorporate the dynamic behaviour of the solids. Correspondingly, the measurementof chemical concentration as a function of particle size must be made for themodel to be useful.

The model to describethe unsteady state change in aqueous phase chemical concentration in the absenceof evaporation, assuming the aqueous phase to be well mixed is shown inequations 1a and 1b(1) - 1b(n). Equation 1a represents the change in aqueousphase concentration, which is the sum of contributions of desorptionof chemical from each discrete particle size. Equations 1b(1-n) representchemical balance on each of the 'n' particle fractions of size dp_i .The desorption kinetics are dependent on thesolid-water mass transfer coefficient, k^2_{A3}. The driving force is the concentrationdifference between the aqueous concentration and the concentration inequilibrium with the chemical present on the solids, where K_{A32} isthe solid-water partition constant. The second term in equation 1a representsthe rate of evaporation.

$$\frac{d}{dt}(\rho_{A2}) = \sum_{i=1}^{n} \frac{6\, k^2_{A3}\rho_{33i}}{\rho_p d_{pi}} \cdot \left[\frac{(\rho_{33i} W_{3i})}{\rho_{33i} K_{A32}} - \rho_{A2}\right] - K_{A}(\rho_{A2} - \rho^*_{A2}) \quad (1a)$$

$$\frac{d}{dt}(\rho_{ss}W_s)_i = -\frac{6\,k_{AB}^2\rho_{ss,i}}{\rho_p d_{p,i}}\cdot\left[\frac{(\rho_{ss,i}W_{s,i})}{\rho_{ss,i}K_{AB2}} - \rho_{AD}\right] \quad \text{(1b-1to 1b-n)}$$

The quantity $\rho_{ss,i}$ is the suspended solidconcentration that is dependent on the particle size and changes with time. Inthe model, it is assumed that the suspended solids concentration changes as aresult of the settling. The equationswere solved using Matlab along with the estimate of $\rho_{ss,i}$ at each time step. Thismodel is based on an earlier version of the model which did not consider thetime-dependency of suspended solids concentrations or the particle sizedependent loading [2].

One of the primary inputs to the model described above isthe particle size dependent distribution of contaminant concentration (loading)in the solids. An experimental technique based on the separation of differentfractions based on settling is also described here. Tests were conducted toevaluate the efficiency of separation and the different fractions. In thesetests, the solids, chemical and organic carbon balance was measured to validatethe separation process. The separation of solids using the settling method isreasonably effective with a solids balance that is satisfactory. Relativeloading of different size fractions were collected and were different. Themodel was tested for conditions and the performance was evaluated withreference to the previous model as well as experimental data obtained forchemical desorption.

See more of this Session: Contaminant Transport and Site Remediation
See more of this Group/Topical: Environmental Division

(50c) Application of Supported Liquid Membrane (SLM) Technology In Recycling of Li+ From Battery Scrap Leachate

Hannes Noll, Marlene Fritz and Matthaeus Siebenhofer, Institute of Chemical Engineering and Environmental Technology, Graz University of Technology, Graz, Austria

In the last years the usage of Lithium Ion Batteries (LIB) has notably increased. LIB are used as power sources in cell phones, laptops, video cameras etc. As a consequence the demand for recycling processes for the valuable components of spent Lithium ion batteries, such as Cobalt, Copper, Nickel, Lithium etc. is highly desirable. Battery Scrap from Lithium Ion Batteries is treated by acid or alkaline leaching to get the target constituents dissolved. Leaching solutions typically contain Co^{2+}, Al^{3+}, Cu^{2+} and Li^{+} ions. Supported Liquid Membrane Technology may serve as unit operation for selectively separating the constituents. .

The aim of this work was to develop complete recycling processes for Lithium recycling from aqueous solutions by implementation of Supported Liquid Membrane Technology (SLMT).

Pertraction of Cobalt, Copper, Aluminum and Lithium through the Ion Exchangers DEHPA and LIX 54 was investigated in order to evaluate whether selective separation of the constituents Cobalt, Copper and Aluminum from Li^{+} by SLMT is feasible.

Microporous Polyethylene membranes with a thickness of 2 mm and a pore size distribution of 7-12 μm served as membrane supports. The pertraction experiments were conducted in a lab scale unit, providing two chambers for the aqueous donor and acceptor solutions, separated by the porous membrane support containing the extractants. Different flux of the metal ions of copper and cobalt into the acceptor solution was applied as selectivity criterion, leaving Li^{+} finally in the raffinate phase for further processing.

See more of this Session: Reactive Extractions
See more of this Group/Topical: Separations Division

(50b) Design Optimisation In Liquid Membrane Permeation with Supported Membranes

Marlene Fritz[1], Hannes Noll[1] and **Matthus Siebenhofer**[2], (1)Institute of Chemical Engineering and Environmental Technology, Graz University of Technology, Graz, Austria, (2)Institute of Chemical Engineering and Environmental Technology, Graz University of Technology, 8010 Graz, Austria

Liquid membrane permeation with support layers is a promising technology for heavy metal recycling as well as applications in the chemical industry, food industry and pharmaceutical industry.

Goal of this research project is the optimisation of design for industrial applications.

After having confirmed the design principle of filter press arrangement scale up and operation in continuous mode was commissioned. Operation results of bench scale operation were confirmed after scale up.

In the next step optimisation of hydrodynamics of the inlet of the feed- and stripping phase was carried out. Due to laminar flow in the feed phase as well as in the stripping phase the residence time distribution may affect mass transfer performance. To obtain a narrow RTD pattern the compartments were therefore equipped with flow rectifiers.

Finally membrane stability was tested with two types of support layers. With a single-layer support long term stability test was abandoned after 2000 hours. With a triple-layer support long term stability was tested up to 7000 hours.

See more of this Session: Reactive Extractions
See more of this Group/Topical: Separations Division

(563a) Response of the UO2 (111) Surface to Varying Oxygen Partial Pressure

Gloria A. E. Oxford and Anne M. Chaka, Physical Measurement Laboratory, National Institute of Standards and Technology, Gaithersburg, MD

UO_2corrosion is a significant concern for the nuclear power industry as well asfor the weapons industry. Oxidation ofUO_2 can lead to significant expansion of the material, which inturn, can cause cracking and spallation.While limited studies of oxidation and reactivity of bulk $UO_2$and the UO_2 (111) surface have been performed using periodic densityfunctional theory (DFT), surface and subsurface oxidation as a function oftemperature and pressure has not been explored.We present periodic DFT calculations of the clean and oxidized UO_2(111) surfaces with and without subsurface oxidation. Generally, the generalized gradientapproximation (GGA) formulation used in this work gives highly exothermicoxidation energies, which is consistent with experimental data and in contrastto previous GGA+*U* studies,demonstrating that GGA describes the energetics of this system better. Using *abinitio* thermodynamics, the temperatures and pressures needed to oxidize thesurface and perform initial subsurface oxidation have been calculated. We also analyze the uranium magnetic andelectronic structure as well as local geometry in response to oxidation.

See more of this Session: Advances In Numerical Simulations Bridging Chemical and Nuclear Engineering Phenomena or Processes
See more of this Group/Topical: Nuclear Engineering Division

(563d) Adsorption of Hard Spheres: Development of a Re-Summed Bridge Function Theory for the Structures and Work of Insertion At Moderate to High Densities (ρ0 = 0. 5745 ~ 0.9135)

Lloyd L. Lee, Department of Chemical and Materials Engineering, The California State University Pomona, Pomona, CA and Giuseppe Pellicane, School of Physics, University of Kwazulu-Natal, 3209 Pietermaritzburg, South Africa

We propose a star-function bases density functional theory* for accurate determination of the density profiles in the hard sphere/hard wall adsorptive system. We transform the Euler-Lagrange equation into an equivalent bridge-function based equation. The bridge function is analyzed in terms of a functional Taylor expansion and the terms in the infinite expansion are re-summed into a closed expression** for calculation. This new closure is tested on the simple system of hard spheres. Results for bulk densities varying from $\rho_0 = 0.\ 5745 \sim 0.9135$ are compared with Monte Carlo data and very close agreement is obtained. The exact sum rule of wall density is incorporated into the closure. We also characterize the insertion work function *W(z)* (work required to insert a test particle into the fluid at position *z* away from the wall) as the density varies. Comparison is made with other theories on the hard sphere fluids.

*.. Lee L. L., J. Chem. Eng. Data, 2010, 55, 1897.

.. Lee L. L., Pellicane G., Chapman W.G., J. Supercritical Fluids 2010, **55, 524.

See more of this Session: Advances In Numerical Simulations Bridging Chemical and Nuclear Engineering Phenomena or Processes
See more of this Group/Topical: Nuclear Engineering Division

Kinetics of Self-Hydrolysis of Concentrated Sodium Borohydride Solutions at High Temperatures

Rajasree Retnamma[1], Lin Yu[2], Carmen M. Rangel[3], Augusto Q. Novais[1], Karl Johnson[4], Michael A. Matthews[2]

[1]Energy Systems Modeling and Optimization Unit, National Laboratory of Energy and Geology, Lisbon, Portugal
[2]Department of Chemical Engineering, University of South Carolina, USA
[3]Fuel Cells and Hydrogen Unit, National Laboratory of Energy and Geology, Lisbon, Portugal
[4]Department of Chemical and Petroleum Engineering, University of Pittsburgh, USA

1. Introduction

Hydrolysis of sodium borohydride ($NaBH_4$) solution over metal catalyst is accepted as a potential technology for delivering H_2 to portable fuel cells. However, unresolved issues such as minimizing the amount of water and the nature of hydration of metaborate by-products limit its hydrogen storage capacities. An alternative method, steam/water vapor hydrolysis, can enhance the hydrogen storage capacities without catalyst, if operating conditions can be optimized [1]. In this approach, solid $NaBH_4$ absorbs water and deliquesces, forming a highly concentrated viscous solution at near boiling point of water, followed by hydrogen evolution from the concentrated viscous solution. It is also well understood that self-hydrolysis occurs even at room temperature when $NaBH_4$ is mixed with water and the rate becomes significant at high temperatures. Self-hydrolysis must be arrested to increase the shelf-life of the solution.

The knowledge of the kinetics of self-hydrolysis of concentrated solutions at elevated temperature is thus crucial for: (1) developing steam/water vapor hydrolysis technology; and (2) handling and storage of $NaBH_4$ solutions. Concentrated solutions have not been studied extensively so far and the reported studies, except two [2, 3], were for buffered dilute solutions of $NaBH_4$ (< 0.4 wt%) at low temperatures (< 25 ^{0}C). The present study reports the kinetics of self-hydrolysis of unbuffered, unstabilized concentrated $NaBH_4$ solutions (10-20 wt%) at high temperatures (25-80 ^{0}C) by a differential isothermal isoconversional method.

2. Theoretical

The $NaBH_4$ hydrolysis (self or catalytic) reaction occurs according to [4]:

$$NaBH_4\text{ (aq or s)} + (2+x)\, H_2O\text{ (l)} \longrightarrow NaBO_2.\ x\, H_2O\text{ (aq or s)} + \text{heat} \qquad (1)$$

where x is the excess hydration factor.

The mechanism of this process is highly complex and too complicated to be described by a kinetic model within the practical operating conditions [5]. The isoconversional method is often used to describe such kinetics, where, assuming a single step kinetic approximation, the rate expression is [6].

$$\frac{d\alpha}{dt} = k'_{ap}(T, C_0)\, f(\alpha) \qquad (2)$$

$k'_{ap}(T, C_0)$ is the apparent rate constant, which depends on temperature T and initial reactant concentration C_0, α is the extent of conversion and $f(\alpha)$ the conversion function representing the mechanism of the process.

The $NaBH_4$ hydrolysis process exhibits maximum rate at the beginning of the reaction, which decreases continuously with the extent of conversion. The most appropriate form of $f(\alpha)$ for such decelerating type process is the 'reaction order model' [7]:

$$f(\alpha) = (1 - \alpha)^m \qquad (3)$$

where m is the reaction order.

Combining Eqs. (2)- (3), and considering the definition of extent of reaction α gives:

$$\frac{d\alpha}{dt} = k_{ap}(T)\, C_0^{m-1}\, (1 - \alpha)^m \qquad (4)$$

where,

$$k'_{ap}(T, C_0) = k_{ap}(T) * C_0^{\,m-1} \qquad (5)$$

For a fixed conversion (α), Eq. (4) reduces to

$$\left(\frac{d\alpha}{dt}\right)_\alpha = k''_{ap}(T, \alpha)\, C_0^{m-1} \qquad (6)$$

where,

$$k''_{ap}(T, \alpha) = k_{ap}(T) * (1 - \alpha)^m \qquad (7)$$

Equation (6), after a logarithmic transformation:

$$\ln\left(\frac{d\alpha}{dt}\right)_\alpha = \ln k''_{ap}(T, \alpha) + (m - 1)\ln C_0 \qquad (8)$$

For an isothermal process, the values of $\ln k''_{ap}(T, \alpha)$ and m are determined by Eq. (8) for each temperature and α, for different initial borohydride concentrations.

The temperature dependence of the rate constant $k_{ap}(T)$ is then expressed by Arrhenius relationship:

$$k_{ap}(T) = A_{ap} \exp\left(\frac{-E_{ap}}{RT}\right) \quad (9)$$

The apparent activation parameters, i.e. apparent activation energy (E_{ap}) and pre-exponential factor (A_{ap}) are estimated by linear regression of ln $k_{ap}(T)$ vs. 1/T data.

3. Results and discussion

3.1. Kinetic study

Effect of $NaBH_4$ conversion and temperature on reaction rate

The experimental kinetic data used were reported by Yu and Matthews [8], in which the evolution of metaborate formation and $NaBH_4$ consumption were generated *in-situ* by ^{11}B NMR technique for unbuffered, unstabilized 10, 15 and 20 wt% $NaBH_4$ solutions at 25, 60 and 80 ^{0}C. The estimated reaction rates from the experimental data for various $NaBH_4$ conversion, initial $NaBH_4$ concentration and temperature are given in Table 1. As evident, the reaction rate increases with both temperature and initial $NaBH_4$ concentration for a fixed $NaBH_4$ conversion and decreases with increase in $NaBH_4$ conversion for a fixed temperature and initial $NaBH_4$ concentration. Table 1 is then exploited to determine the reaction order and apparent activation parameters with respect to $NaBH_4$ conversion and temperature.

Table 1. Reaction rates for different $NaBH_4$ conversions at various temperatures and initial $NaBH_4$ concentrations.

$NaBH_4$ (wt %)	T(^{0}C)	Reaction rates (r, mol L^{-1} hr^{-1}) at various $NaBH_4$ conversions									
		5 %	10 %	15 %	20 %	25%	30 %	35 %	40 %	45%	50%
10		0.086	0.048	0.025	0.015						
15	25	0.136	0.080	0.037	0.027						
20		0.241	0.108	0.057	0.036						
10		0.637	0.559	0.485	0.415	0.345	0.273	0.210	0.158	0.110	0.100
15	60	0.740	0.654	0.569	0.483	0.398	0.317	0.242	0.181	0.144	0.130
20		0.897	0.797	0.697	0.598	0.499	0.402	0.310	0.227	0.160	0.143
10		2.209	2.089	1.970	1.848	1.723	1.597	1.468	1.340	1.201	1.062
15	80	2.384	2.241	2.130	1.981	1.862	1.720	1.687	1.470	1.352	1.232
20		2.698	2.555	2.410	2.264	2.115	1.964	1.810	1.653	1.493	1.326

Effect of $NaBH_4$ conversion and temperature on reaction order

Fig. 1 shows the reaction order as estimated by linear regression of rate versus $NaBH_4$ concentration for various $NaBH_4$ conversions and. As seen, the reaction order remains constant at a given temperature irrespective of $NaBH_4$ conversion, while it decreases with increase in temperature irrespective of $NaBH_4$ conversion. The order decreases from first- to 0.25 with increase in temperature from 25 to 80 ^{0}C. The estimated rate constant, $k_{ap}(T)$ (not given), decreases with increase in $NaBH_4$ conversion at constant temperature and increases with temperature at constant $NaBH_4$ conversion. It is noteworthy that the variation of reaction order with temperature has been reported by Vyazovkin and Sbirrazzuolui [9] for complex epoxy-anhydride cure kinetics.

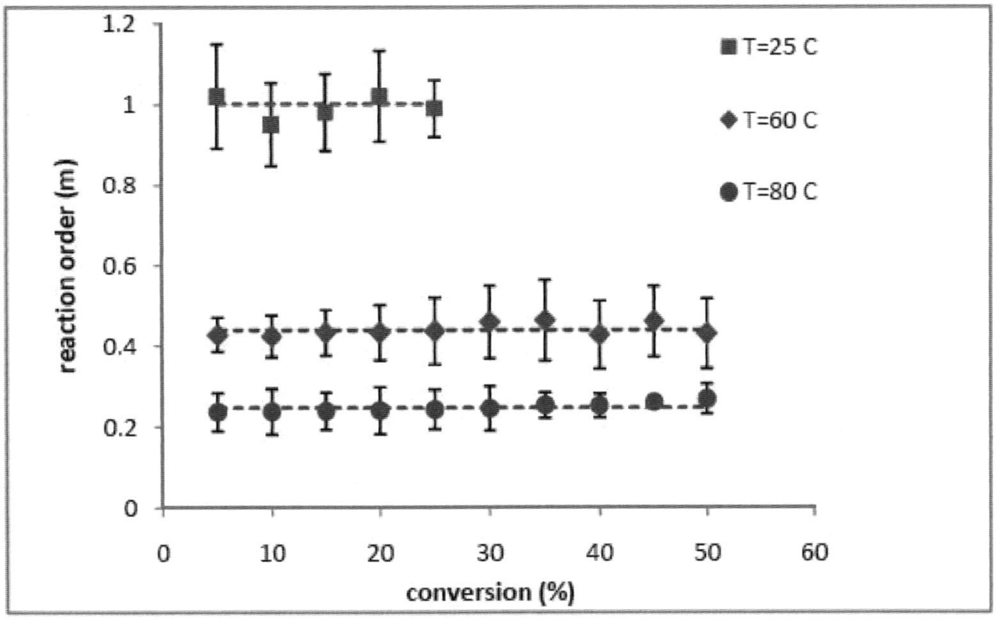

Figure 1. Effect of $NaBH_4$ conversion (%) on reaction order

It is well known that the solution viscosity increases during the course of hydrolysis reaction and this has been considered the reason for the observed decrease in rate with reaction time or with $NaBH_4$ conversion [5]. The effect of temperature on solution viscosity can be described in terms of Arrhenius-type relationships [10]:

$$\eta = \eta_0 \exp\left(\frac{E_{af}}{RT}\right) \qquad (10)$$

where η is the viscosity, η_0 the pre-exponential factor and E_{af} the activation energy of flow. This implies that increase in temperature decreases the solution viscosity. Viscosity is an indication of the solution resistance to flow, and reduced flow resistance at high temperature enhances the interaction between reactant molecules. This in turn lessens the relative contribution of the effect of the concentration on the reaction rate. The deviation from first-order kinetics at low temperatures to fractional-order kinetics at higher temperatures observed in the present study, is attributed to this difference in interaction between reactant molecules with respect to temperature due to change in solution viscosity. It is worthwhile noting that the earlier studies also reported first-order kinetics for self-hydrolysis of $NaBH_4$ for lower conversions and temperatures and for buffered solutions [11-15]

Effect of $NaBH_4$ conversion on activation energy

The classical kinetic theory defines reaction orders to be constant with temperature. In the present case the observed difference in the reaction order with temperature (Fig. 1) suggests a $NaBH_4$ conversion effect on the activation energy of the hydrolysis reaction (E_a), as given in Table 2. As seen, activation energy increases with $NaBH_4$ conversion/with formation of metaborate by-products.

It is well documented in literature that the formation of highly basic metaborate by-products in the course of the hydrolysis reaction results in highly viscous solutions. Investigation on the influence of soluble solids content on viscosity, by various researchers, showed that depending on the type of soluble solids content the activation energy of flow (E_{af}) might increase or decrease. The effects of soluble solid contents are described by the following expressions [10]:

$$\eta = \eta_1 C^{b_1} \quad (11)$$

$$\eta = \eta_2 \exp(b_2 C) \quad (12)$$

where η_i and b_i are constants and C is the concentration. Hence one may assume that the activation energy of the flow increases with an increase in $NaBH_4$ conversion due to metaborate by-product formation, which in turn increases the activation energy of the hydrolysis reaction.

Table 2. Apparent activation energy determined for various $NaBH_4$ conversions

$NaBH_4$ Conversion (%)	[a]E_{ap} (kJ mol^{-1})	R^2	Standard error	F	P-value
5	64. 43 ± 14.5	0.9944	1.14	3914.89	0.025
10	71.03 ±11.5	0.9998	0.91	6721.39	0.038
15	82.50 ± 13.5	0.9998	1.03	6725.19	0.008
20	89.62 ± 12.5	0.9994	1.96	2152.88	0.013

[a]with ± 95% confidence interval

3.2. Modeling

The dependence of apparent activation energy (E_{ap}) on α (Table 2) is described by [16]:

$$E_{ap} = \varepsilon_0 + \varepsilon_1 \ln(1-\alpha) \quad (13)$$

with ε_0 = 54.2 ± 1.7 kJ mol^{-1} and ε_1 = -114±12.2 kJ mol^{-1} (within a 95% confidence interval)

The change in apparent reaction order with temperature (Fig. 1) can be evaluated if the parameters of the kinetic compensation effect (KCE) are known [16]. The present study shows that $NaBH_4$ hydrolysis follows the following KCE relationship:

$$\ln A_{ap} = \alpha^* + \beta^* E_{ap} \qquad (14)$$

with α^* =0.84 ± 1.2 (mol $L^{-1})^{1-(m+n)}$ hr^{-1} and β^*= 0.36 ± 0.04 kJ mol^{-1}(within a 95% confidence interval)

Combining Equations (4), (9), (13) and (14) results in:

$$\frac{d\alpha}{dt} = A_0(1-\alpha)^{(m+n)} \qquad (15)$$

where:

$$n = \beta^* \varepsilon_1 - \frac{\varepsilon_1}{RT} \qquad (16)$$

$$A_0 = \exp\left(\alpha^* + \beta^* \varepsilon_0 - \frac{\varepsilon_0}{RT}\right) C_0^{m+n-1} \qquad (17)$$

Equation (15) shows that the overall process is apparently described with a reaction order n+m as illustrated in Fig 1. If the values of n are estimated from equation (16) and n+m from Fig 1, the values for m obtained are given in Table 3. As expected, m tends to a constant value, i.e. 0.22 ± 0.01.

Table 3. Values of n, n+m and m at different temperatures

$T(^0C)$	$n^{(1)}$ (Eq. 16)	$n+m^{(2)}$ (Fig. 1)	$m^{(3,*)}$
25	0.820	1.04	0.214
60	0.222	0.44	0.218
80	0.024	0.245	0.220

[1]dependent on temperature, [2]apparent reaction order
[3] independent on temperature, [*] m = 0.22± 0.01 (mean ± standard deviation)

4. Conclusions

The complex $NaBH_4$ self-hydrolysis process is analyzed by a differential isothermal isoconversional method, based on a single-step kinetic approximation. The overall process is apparently described by a 'reaction order' model with temperature dependent reaction order in a wide practical operating window (10-20 wt% $NaBH_4$ solution and 25-80 ^0C).

The apparent activation parameters, i.e. apparent activation energy and pre-exponential factor are conversion dependent and interrelated through the kinetic compensation effect, which explains the observed effect of the temperature on the reaction order.

The apparent activation parameters obtained provide some insight into the theoretical kinetic mechanism, which however remains a matter of heated debates.

5. References

[1] Beaird AM, Davis TA, Matthews MA. Deliquescence in the hydrolysis of sodium borohydride by water vapor, Ind Eng Chem Res 2010;49(20):9596-9599.

[2] Gonçalves, A., Castro, P., Novais, A.Q., Fernandes, V., Rangel, C. and Matos, H., Dynamic modeling of hydrogen generation via hydrolysis of sodium borohydride, In Proceedings of PRES'07, 24-28 June, Ischia, Italy, (In Chemical Engineering Transactions, Vol. 12, 2007. Editor: Jiri Klemes, AIDIC, Italy, 243.)

[3] Andrieux J, Demirci UB, Hannauer J, Gervais C, Goutaudier C, Miele P. Spontaneous hydrolysis of sodium borohydride in harsh conditions. Int J Hydrogen Energy 2011;36:224-233.

[4] Cavaliere S, Hannauer J, Demirci UB, Akdim O, Miele P. Ex situ characterization of N_2H_4-, $NaBH_4$- and NH_3BH_3- reduced cobalt catalysts in $NaBH_4$ hydrolysis. Catal Today 2011;170:3-12.

[5] Retnamma R, Novais AQ, Rangel C.M. Kinetics of hydrolysis of sodium borohydride for hydrogen production in fuel cell applications: A review. Int J Hydrogen Energy 2011; 36: 9772-9790 (and references thereof).

[6] Simon P. Isoconversional methods: Fundamentals, meaning and application. J. Therm. Anal. Calorimetry 2004;76:123-132.

[7] Vyazovkin S, Burnham AK, Criado JM, Perez-Maqueda AA, Popescu C, Sbirrazzuoli N. ICTAC Kinetics committee recommendations for performing kinetic computations on thermal analysis data. Therm Chim Acta 2011;520:1-9.

[8] Yu Lin, Matthews MA. Hydrolysis of sodium borohydride in concentrated aqueous solution. Int J Hydrogen Energy 2011;36:7416-7422.

[9] Vyazovkin S, Sbirrazzuoli N. Kinetic methods to study isothermal and nonisothermal epoxy-anhydride cure. Macromol Chem Phys 1991;200:2294-2303.

[10] Ibarz A, Garvin A, Costa J. Rheological behavior of sole (prunus spinosa) fruit juices. J Food Engng 1996;27:423-430.

[11] Pecsok R. Polarographic studies on the oxidation and hydrolysis of sodium borohydride. J Ame Chem Soc 1953;75:2862-2864.

[12] Stockmayer WH, Miller RR, Zeto RJ. Kinetics of borohydride hydrolysis. J. Phys. Chem 1961;65:1076-1077.

[13] Davis RE, Swain CG. General acid catalysis of the hydrolysis of sodium borohydride. J Ame Chem Soc 1960;82(22):5949-5950.

[14] Davis RE. Bromels E, Kibby CL. Borohydrides III. Hydrolysis of sodium borohydride in aqueous solution. J Ame Chem Soc 1962;84:885-889.

[15] Mesmer RE, Jolly WL. The hydrolysis of aqueous hydroborate. Inorg Chem 1962;1:608-612.
[16] Budrugeac P, Popescu C, Segal E. Change of the apparent reaction order with temperature as a consequence of the reaction mechanism complexity. J. Therm. Anal. Calorimetry 2001;64:821-827.

(213f) BP Oil Ganglia Transport In Transparent Porous Media At Low Capillary Number

Peixi Zhu and Kyriakos Papadopoulos, Chemical and Biomolecular Engineering Department, Tulane University, New Orleans, LA

Usingvideo-capillary-microscopy, microscopic packed beds were constructed insidecapillaries, using cryolite as the porousmedium. When such packed bed was filled with an aqueous surfactantsolution, the cryolite was renderedtransparent. Oil droplets were injected inside or next to theporous media and their flow/mobilization was monitored. Pressure-drivenflow was induced at varying pressure gradients and at different concentrationsof key surfactants. At low capillary number, e.g., superficial flowrate on the order of 10^{-2}ml/min, the motion of oil phase is mainlydriven by capillary force. The change of the oil phase from ganglia tosmall droplets of a size comparable to that of the pores,was captured in film. Subsequent analysis of the multiphase flow revealedthe evolution of the flow patterns along the axis of the capillary. Thetalk will show videos of the oil mobilization phenomena and will present theconclusions of this on-going research to date.

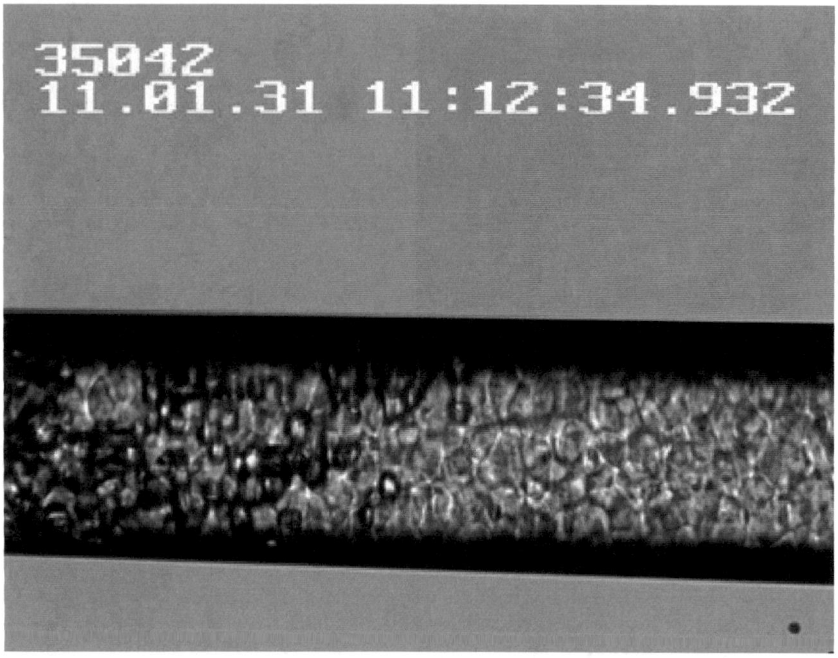

See more of this Session: Contaminant Transport and Site Remediation
See more of this Group/Topical: Environmental Division

Developing Experimental and Computational Nuclear Forensics Signals of Fast Neutron Irradiation Facility at TRIGA Reactor

Ian Schwerdt, Joseph Levinthal, Chris Dances, Luis Ortiz-R,
Dong-OK Choe and Tatjana Jevremovic
University of Utah Nuclear Engineering Program (UNEP)
The University of Utah, Salt lake City, UT

INTRODUCTION

Over 230 research reactors are operating on campuses and labs across the globe. These research reactors are diverse in their design, fuel type, power, and fuel enrichment. Most TRIGA reactors use Uranium Zirconium Hydride (UZrH) fuel that initially was designed to contain High-Enriched Uranium (HEU). A majority have been converted into low-enriched uranium (LEU) fuel (General Atomics)[1]. Fast neutrons can be used in several applications such as neutron hardness experiments for electronics and material tests. Accurate neutron flux and radiation dose measurements and computational estimates in reactors' irradiation facilities are of great importance for producing accurate nuclear signals of interest to nuclear forensics.

The University of Utah houses a 100 kW TRIGA reactor (UUTR), with a few irradiators, the Fast Neutron Irradiation Facility (FNIF), thermal irradiation facility (TI) central irradiation facility. The function of the FNIF is to provide a location for samples to be irradiated by fast neutrons coming from the reactor core. The FNIF is located directly next to the exterior of the reactor core. It consists of an aluminum shell filled with stacked lead bricks. The lead bricks are placed along the walls of the aluminum structure leaving a sizeable gap in the center for samples to be placed. The lead bricks block gamma radiation from the reactor core, while minimizing the amount of energy lost from scattered neutrons. The initial amount of energy of a neutron produced by fission follows a predictable curve called the neutron fission spectrum. The energy spectrum is a bell shaped curve that has a peak at about 1 MeV, and an average value of 2 MeV. As neutrons are produced in the reactor core, they lose energy by interacting with the matter in between the emission point and the FNIF. The vast majority of incident neutrons are produced from the two closest rows of fuel elements in the reactor core. The neutrons are emitted at a random trajectory from the fuel element, and form a cosine flux distribution in each axis direction. The profile within the FNIF is sum of the profiles from the first two rows. This project is largely based on the work presented at the Advancing Tools and Solutions for Nuclear Material Detection 2nd annual national conference by the authors. One of the goals of the previous project was to characterize this distribution of the neutrons with respect to energy and position by using a combination of experimental data and computer simulation. The objective of this project was to refine the simulation, and expand to new experimental techniques.

DESCRIPTION OF EXPERIMENTAL PROCEDURE

One way to map the neutron flux distribution is through the neutron activation of a material and measuring the induced radioactivity, which can be used to develop a profile of the fast neutron flux. For our experiment, we utilized the activation of Sulfur-32 to Phosphorous-32 through a charged particle emission (n, p) reaction. When a moderately fast neutron is captured by the nucleus of Sulfur-32, the nucleus will then emit a proton to become Phosphorous-32. The charged particle emission reaction (n, p) of Sulfur-32 is a threshold reaction requiring a fast neutron with energy in excess of 2 MeV. The Phosphorous-32 has one mechanism of decay, the emission of a detectable beta particle that is used to quantify the amount of neutron flux experienced while in the FNIF.

84 Sulfur-32 pellets were arranged in a 20 cm x 31cm grid pattern and irradiated in the FNIF for 15 minutes at 1kW (Fig. 1. a)). The samples were allowed to decay for seven (7) days so that the activity was low enough for safe handling. The activity of the samples was measured utilizing a Geiger-Mueller tube. The activities were standardized to a reference time corresponding to when the first sample was measured. A general fast neutron flux profile was obtained by plotting the intensity of the activities on a contour map (Fig. 1. b)). The FNIF was then modeled in GEANT4[2], in order to verify that physics modeling could replicate the general spatial variations of the fast neutron flux profile elucidated from the neutron activation of Sulfur-32.

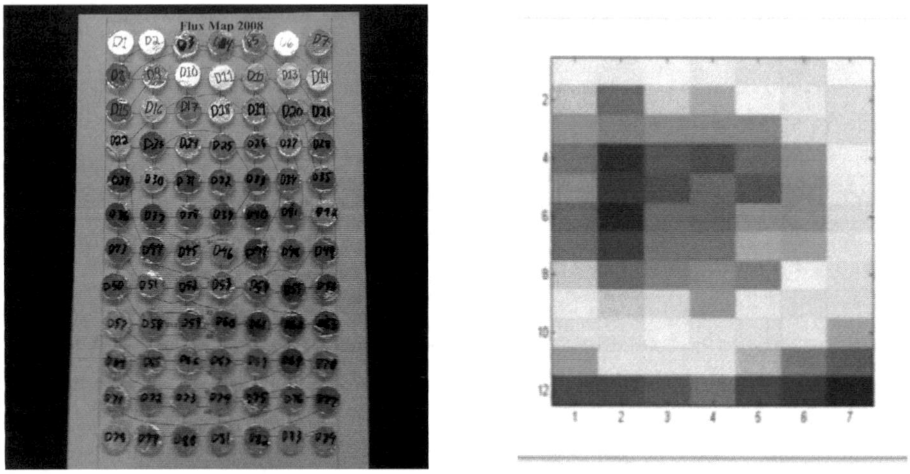

a) b)

Figure 1. a) Sulfur pellets laid out into a grid before irradiation; b) Experimental measurements

In addition to this initial neutron flux mapping of the FNIF, a separate experiment may be able to contribute to this project. A rhodium detector was used to measure the amount of neutron flux in the FNIF. The detector was lowered into the FNIF, and a flux value was determined based on the amperage induced in the detector. An initial flux of 4.6×10^{10} n/cm^{2}sec, was determined in this way (Fig. 2). This rough value was independent of the position of the detector. The value read by the detector represents a fraction of the total flux in the FNIF, at an arbitrary position. Future work with this experiment would be in integrating the detector into the GEANT4 model to aide in benchmarking the detector. The total neutron flux in the FNIF could be scaled from the neutron flux distribution knowing the value of the detector at its position. The FNIF facility offers an opportunity to benchmark the detector, and the computer simulation would aide in confirming the benchmark's validity. This potential advantage in understanding a detector ultimately contributes to the nuclear forensic field. These new detectors under development at UNEP will be optimized for fast detection at site of low and high energy neutrons of interest to nuclear forensics.

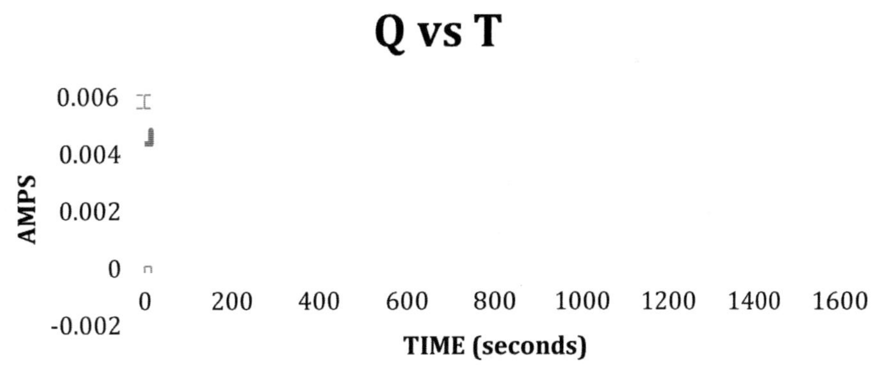

Figure 2. Charge induced in the rhodium detector vs. time to get amperage

DESCRIPTION OF GEANT4 SIMULATION

A computer code developed at CERN called GEometry And Tracking (GEANT4 Collaboration)[2] is used for simulation of FNIF neutron flux distribution. GEANT 4 is an open source code written in C++ designed to model the physics processes involved with particles passing through matter with a wide variety of applications. A simulation is constructed from a number of subclasses and then compiled as a single executable. The most basic simulation requires classes that define the geometry, the particle source, and the physics processes involved. A class called *G4DetectorConstruction* establishes the geometry. After establishing a given shape, it is placed into a *mother* or *world* volume and then is assigned different material properties. For this simulation, a trapezoid constitutes the aluminum casing of the first two rows of the reactor. The FNIF is represented by an aluminum shell encompassing a hollow lead block and is placed next to the aluminum trapezoid (Fig. 3).

A class called *G4particlesource* establishes the generation of any subatomic particles. Each neutron is generated in what is called *an event* with a number of attributes. For each *event* each neutron needs to be assigned a specific initial position, direction, and energy. The position is assigned randomly to a fuel element position in the modeled reactor core. The neutron is then assigned a random position along the surface of the fuel element. The *watt rejection* method is used to assign the neutron an energy level according to the probability of the fission energy spectrum. Finally, each neutron was given a random direction in all three dimensions. A standard physics library was used to simulate the neutron transport established in the physics library class. Additionally, a *scoring method* was used to count the total number of neutrons that pass through the inner chamber of the FNIF. A matrix of cells was established in all three dimensions, and at different energy levels. Each cell then counts the total number of neutrons that pass through indexed cells. From this data, a flux value is assigned to each cell at each energy level. This produces a three dimensional map of the flux at the FNIF and at three different energy levels.

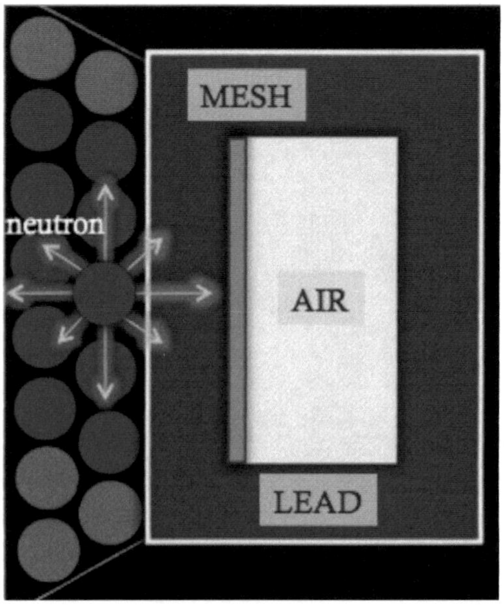

Figure 3. FNIF as modeled in GEANT4. The red block represents the stacked lead bricks in the FNIF. The yellow box in the middle represents the chamber where samples can be placed. The blue circles represent fuel elements, and the green empty slots. As in the diagram, neutrons are generated in the fuel elements and then enter the FNIF.

The results obtained from the simulation were then run through Matlab to graphically display the flux distribution. The three-dimensional flux distribution can be viewed from the top as in Fig. 3, or from the side or front of the FNIF. The side view would be the in the direction from the bottom of Fig. 3. The front view would be from the direction to the right of the image. Each perspective gives the neutron flux distribution along two different axis, leaving a third free. The intensity of the flux distribution on one axis, and the two other represented the cross section image at a particular distance along the third position axis (Fig 4.a). The same could be done with the different energy levels. Instead of altering the third position axis, one could vary the energy range of the distribution. However the profile of the cross sections varies only in the magnitude with respect to energy level. Instead, the energy spectrum inside the FNIF is shown in Fig 4.b). The graph in the top right of the image of Fig. 4. a) and the experimental graph of the data from Fig. 1.b) both depict a similar trend for the neutron flux distribution. Discrepancies arise from the mesh sizes used in plotting the experimental and simulation data. The differences in resolution between the two graphs adds to a bit distort image when comparing experimental with simulation distributions.

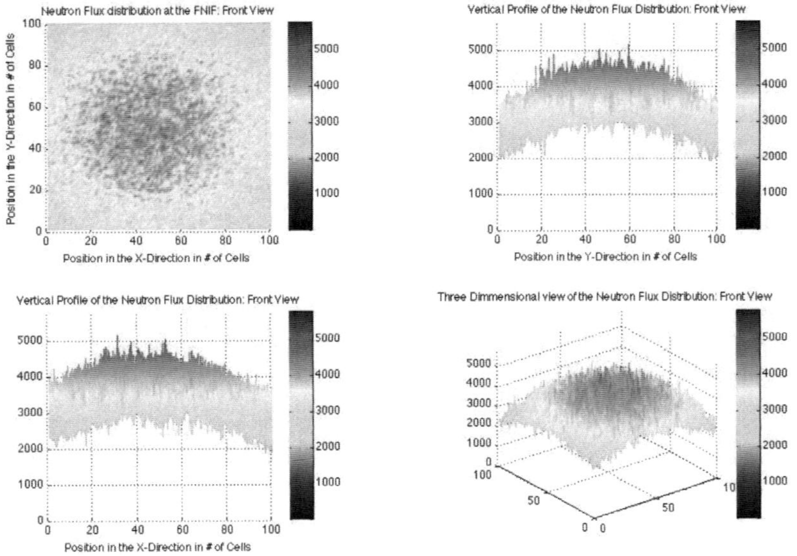

a)

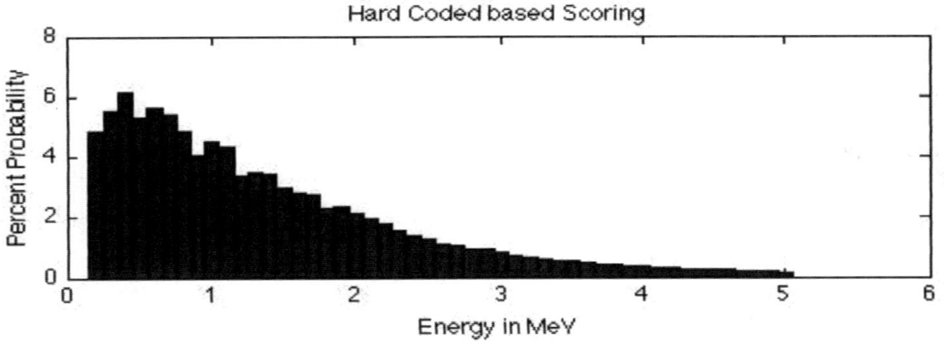

b)

Figure 4. a) The neutron flux distribution from a "forward perspective". That is from the point of view from the reactor towards the FNIF. The flux relative to both the x and y dimensions is shown. The other views are from other angles, of the same 3-D flux map. This same map is one cross section within the facility, and can be taken for any direction; b) The total energy spectrum in the FNIF. While the shape of the curve deeply resembles the fast fission spectrum, the peak has dropped from 2 MeV to just under .5 MeV. This shift also moves the average closer to approximately 1 MeV.

CONCLUSION

Recent development of simulation and experimental models regarding beam ports (irradiation facilities) at the UUTR includes a number of undergraduate and graduate students in a team work. With this effort we are now systematically analyzing each of the irradiation facilities in recording experimental measurements used for calibrating radiation detectors and benchmarking GEANT4 as well as MCNP5/X simulation models. In this paper we presented neutron flux distribution in one of UUTR irradiation ports that was mapped experimentally and modeled using GEANT4. Experiment was based on using the sulfur tablets that where placed on a grid, and then irradiated; the activity of the tablets was then measured. While the coarse grid was used, the experimental data provided accurate cosine flux distribution. This trend was shown to coincide with the computer simulation in GEANT4 reasonably well. GEANT4 simulation provided a higher resolution of the neutron flux distribution and also suggests an energy spectrum within the FNIF. The first measurements using rhodium detector has provided initial data for further comparisons. Future work would lie in integrating the rhodium detector measurements with GEANT4 simulations; this comparison will provide a better understanding of the detector accuracy.

References

1. General Atomics, "TRIGA Nuclear Reactors".

   Accessed: 9/10/2011 at http://www.ga-esi.com/triga/

2. GEANT4 Collaboration, "GEANT4 User's Application Guide for Application Developers"

   Accessed: 9/10/2011 at

   http://geant4.web.cern.ch/geant4/UserDocumentation/UsersGuides/ForApplicationDeveloper/html/index.html

Monte Carlo Simulation of Tritium Generation in The University of Utah TRIGA Reactor

Todd Sherman, Dong-OK Choe, Haori Yang, and Tatjana Jevremovic

Nuclear Engineering Program

The University of Utah, Salt Lake City, UT

Extended Abstract

INTRODUCTION

The University of Utah Nuclear Engineering Program houses a Mark I TRIGA Reactor, UUTR (University of Utah TRIGA Reactor). This reactor is a pool type reactor with water used as both the coolant and neutron moderator. Regular city water is filtered and then dispensed into the tank. UUTR core uses low enriched uranium fuel.

The UUTR is licensed by the Nuclear Regulatory Commission to operate at 100 kW. However, the usual operating power is 90 kW. At this power level a maximum neutron flux of about 3.6 $\times 10^{12}$ neutrons/cm^2-sec is produced at the center of the core. This flux decreases by about an order of magnitude toward the periphery of the core. **Figure 1** shows the layout of the UUTR, irradiation facilities, and the location of some of the heavy water elements.

The UUTR is equipped with various experimental irradiation facilities for neutron activation, fast neutron irradiation and gamma irradiation experiments, as indicated in **Figure 1**. The UUTR's experiments mainly include investigations of radiation hardening, neutron activation analysis, biomedical analysis and isotope production.

TRITIUM (T) GENERATION IN THE UUTR

Where T is generated in UUTR. In the outer ring of the UUTR core there are heavy water elements serving as both moderator and reflector. Neutrons that have reached the heavy water elements will likely be reflected back towards the core instead of escaping from the core. These heavy-water elements are filled with 68% D_2O and 32% H_2O. The TI (**Figure 1**) is also filled with heavy water of the same concentration. The heavy water elements and the TI are the primary sources of tritium production in the UUTR.

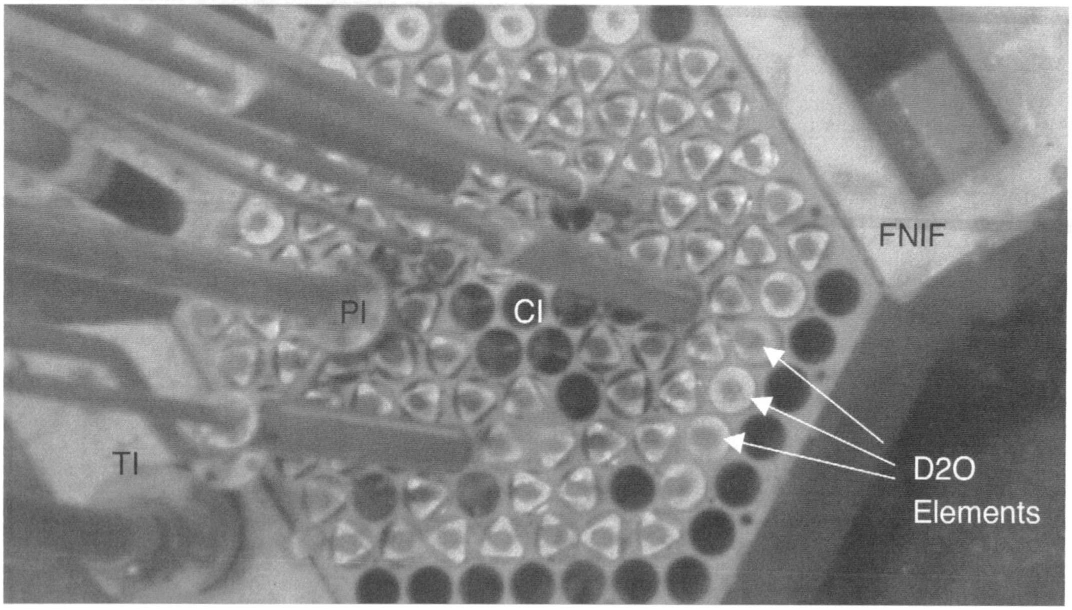

Figure 1 – The University of Utah TRIGA Reactor

[CI – Central Irradiation Facility; TI – Thermal Irradiation Facility; FNIF – Fast Neutron Irradiation Facility; PI – Pneumatic Irradiation Facility]

How is T generated? Hydrogen captures a neutron to become deuterium. Deuterium captures a neutron to become tritium, as illustrated by **Equation 1** and **Figure 2** (Dyadin, 1979). Tritium has a smaller scattering cross section than the other hydrogen isotopes. Tritium is unstable and undergoes a beta decay to become a stable helium-3. The half-life of tritium is 12.33 years.

Why it is important knowing the quantity of generated T?[1] Identifying and managing tritium quantity is important because of its potential adverse characteristics. Tritium, in large doses, can have serious biological consequences, both physically and radiologically. Tritium is also a key component in nuclear weapon proliferation, which makes it important to control. The EPA Safe Drinking Water Act limits the tritium concentration to 0.74 Bq/mL or 20,000 picocurie/ liter. This is the equivalent to 4.15 x10^{11} tritium atoms/ mL in drinking water.

[1] "Tritium is used by all nuclear weapons states to increase the explosive yield of atomic bombs and to miniaturize them. However, this radioactive material has not yet been put under appropriate international control comparable to the nuclear safeguards applied for plutonium and uranium. It is a neglected material in efforts to control the spread and further development of nuclear weapons." (Kalinowski, 2005)

$$ {}_{1}^{1}H + {}_{0}^{1}n \rightarrow {}_{1}^{2}H + {}_{0}^{1}n \rightarrow {}_{1}^{3}H \quad (1) $$

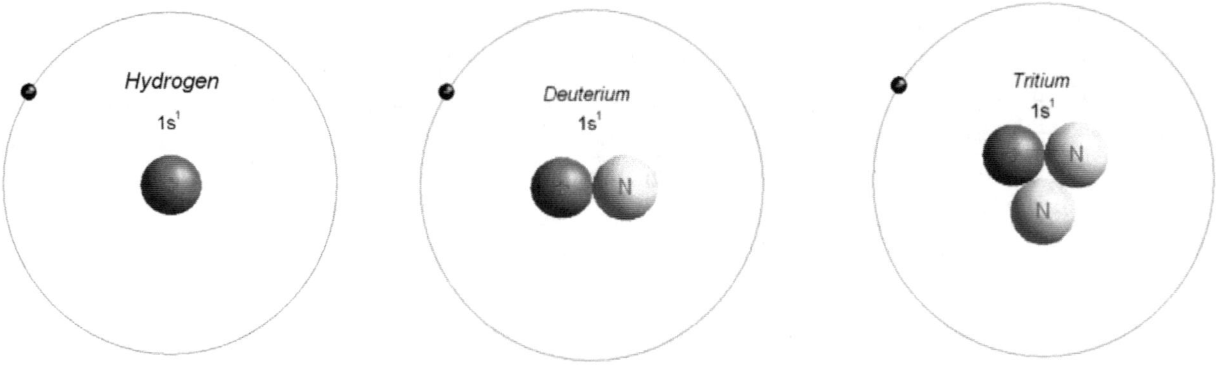

Figure 2 - Three different hydrogen isotopes

ANALYTICAL ESTIMATES OF TRITIUM GENERATION IN UUTR

The tritium generation in several heavy water reactors can be estimated analytically. By using simple neutron activation equations with a measured thermal neutron flux (Erykalov, 2005) we may derive tritium generation based on a very simplified reactor model. The following equation can be used to calculate the tritium generation rate in the UUTR heavy water elements:

$$ N_{Tritium} = N_{deuterium}\,\sigma\phi \quad (2) $$

where:

N = number of atoms

ϕ = neutron flux (neutrons/cm^2-sec)

σ = Thermal neutron absorption cross section (cm^2)

An approximated thermal neutron flux of 3.6 x10^{11} neutrons/cm^2-sec was used to estimate tritium generation in the heavy water elements. The flux in each of the heavy water elements could be measured experimentally; such experiments have not yet been performed but are planed to be completed in 2012. The calculated tritium generation rate using Equation 2 in every heavy water element, at 90 kW, was estimated roughly to be in order of 4.758x10^9 atoms/ sec.

MCNP5 MODELING OF TRITIUM GENERATION

MCNP5 Model. The MCNP5 model of the UUTR core including all details is often used in predicating the outcomes of various experiments and operational settings that would otherwise be difficult to do experimentally. Key parameters (such as neutron flux inside the various irradiation facilities) were already experimentally measured and used to benchmark the MCNP5 model, so that any other modeling of different configurations or the experiments give the known confidence. The measured value of the neutron flux in for example the CI (**Figure 1**) of 3.6 x10^{12} neutrons/cm^2-sec was used to scale the tallied MCNP5 flux values for this model. The CI is where the neutron flux is measured and well known at our facility. So the CI flux values were chosen to scale all the MCNP5 results to the measured values. This scaling factor was produced by dividing the measured flux by the MCNP5 flux tally. The purpose of the scaling factor is to normalize all of the MCNP5 results to same ratios.

The MCNP5 simulation was ran using 100,000 particles, which was enough to bring the relative errors under 0.1, which MCNP5 community considers as reliable. The tritium production was tracked and tallied using the track length estimation. In MCNP5, the average neutron flux in a certain cell can be calculated using track length estimation (F4 tally). In order for tritium to be produced in the UUTR it must come from a deuterium nucleus absorbing a neutron. So the neutron flux is imperative to track and record. The MCNP5 track length estimation tallies the relative average neutron flux. The average neutron flux in a cell can be written as:

$$\phi v = \frac{1}{V}\int dE \int dt \int dV \int d\Omega\, \Psi(\vec{r},\Omega,\mathrm{E},\mathrm{t})$$

$$= \frac{1}{V}\int dE \int dV \int d\Omega \int dt\, vn(\vec{r},\Omega,\mathrm{E},\mathrm{t})$$

$$= \frac{1}{V}\int dE \int dV \int dt\, vN(\vec{r},\mathrm{E},\mathrm{t})$$

The quality $dt \cdot v \cdot N(\vec{r},E,t)$ can be thought of as a track length density. Using F4 tally, MCNP5 estimates the average flux based on the accumulation of weighted track length for all particles tracks in the cell. Based on neutron flux calculated using F4 tally, the FMn card can be used to calculate any quantity of the form

$$C\int \phi(E)R_m(E)dE$$

$R_m(E)$ can be any response function from the MCNP cross-section library. The tritium generation rate in a heavy water element can be written as

$$\int \phi(E)\Sigma(E)dE$$

where $\Sigma(E)$ is the macroscopic cross section for deuterium capture reaction. It can be rearranged as

$$N\int \phi(E)\sigma(E)dE$$

where N is the number density of deuterium. The reaction rate per unit volume is then

$$\frac{N}{V}\int \phi(E)\sigma(E)dE$$

Thus we can set the multiplicative constant on the FMn card as the deuterium density per cm^3

$$C = \frac{N}{V}$$

The MCNP5 F4 tally gives a unit neutron flux, ϕ, per unit time (neutrons/ cm^2). The FMn card simply multiplied ϕ with the microscopic neutron absorption cross section (barns) σ. The FMn output, therefore, was obtained as:

$$\phi\left(\frac{neutrons}{cm^2}\right)\sigma(barns) = \phi\sigma\left(\frac{neutrons}{cm^2}(barns)\right) \quad (3)$$

To convert this value into an actual tritium generation rate per cm^3 the results were multiplied by the scaling factor, the constant C, the conversion from $1x10^{-24}$ cm^2 = 1 barn and the volume of each heavy water element; these are shown in **Table 1**. The following relation explains this final calculation step:

$$SF\left(\frac{\frac{neutrons}{cm^2 \cdot sec}}{\frac{neutrons}{cm^2}}\right)\phi\sigma\left(\frac{neutrons}{cm^2}(barns)\right)C\left(\frac{N_{deuterium}}{cm^3}\right)\left(\frac{1capture}{neutron}\right)\left(\frac{1x\,10^{-24}cm^2}{1\,barn}\right)V(cm^3) = \frac{tritium\ generations}{sec} \quad (4)$$

where:

SF = scaling factor obtained by taking ratio of the measured CI flux and the MCNP5 simulation results
$\phi\,\sigma$ = FMn card results from MCNP5 simulation
C = multiplication constant used on the FMn card in the MCNP5 simulation
V = volume of a heavy water element

Tritium quantities in UUTR obtained by MCNP5. The MCNP5 gives the unit interactions. Thus all the MCNP5 values where multiplied by the scaling factor as well as the heavy water element volume (cm^3) and deuterium atom density (atoms/ cm^3). These results are summarized in **Table 1**. The unit tritium generation activities are also shown in **Table 1** as calculated from:

$$A = \lambda N \quad (5)$$

A = Activity (Bq)

λ = decay constant (sec^{-1})

N = Deuterium density per unit volume (atoms/ cm^{3})

As shown in **Table 1**, the tritium generation rate in twelve heavy water elements at the UUTR is far below the EPA regulation even when the reactor is running at its full licensed thermal power. The tritium generation rate inside the TI is also well below the EPA regulation. Thus, the amount of tritium generated at the UUTR due to normal operation will not pose any radiological hazard to the staff members or the public if by any unexpected event is released to the environment.

Table 1 - Tritium Generation in the UUTR based on MCNP5 Model

Heavy Water Element	MCNP5 Tritium generation rate (Atom/sec)	Unit tritium activity generation rate (Bq/cm^3)
TI	$3.2899 \times 10^{10} \pm 0.0187$	4.77×10^{4}
1	$1.3294 \times 10^{9} \pm 0.0369$	4.21×10^{3}
2	$1.5884 \times 10^{9} \pm 0.0346$	5.03×10^{3}
3	$1.5543 \times 10^{9} \pm 0.0342$	4.92×10^{3}
4	$1.3287 \times 10^{9} \pm 0.0368$	4.21×10^{3}
5	$1.5168 \times 10^{9} \pm 0.0330$	4.80×10^{3}
6	$1.4818 \times 10^{9} \pm 0.0341$	4.69×10^{3}
7	$1.7073 \times 10^{9} \pm 0.0336$	5.41×10^{3}
8	$1.6108 \times 10^{9} \pm 0.0347$	5.10×10^{3}
9	$1.4987 \times 10^{9} \pm 0.0348$	4.75×10^{3}
10	$1.3438 \times 10^{9} \pm 0.036$	4.26×10^{3}
11	$1.5638 \times 10^{9} \pm 0.0338$	4.95×10^{3}
12	$1.3477 \times 10^{9} \pm 0.0363$	4.27×10^{3}

CONCLUSION

The tritium generation inside of the UUTR is investigated using a detailed model of the UUTR based on MCNP5. In order to obtain a tritium generation profile in the existing twelve heavy water elements and the TI, the MCNP5 values are normalized to a single source particle. To get the overall tritium generation rate, a scaling factor was calculated by taking ratio of the measured and simulated neutron flux at the CI (**Figure 1**). The MCNP5 simulation model as presented provided reliable values for tritium generation rate and activity. This model now serves as a base-line for our feasibility study of the UUTR power upgrade.

References

Dyadin, Y. V. (1979). Measurement of tritium concentration in heavy-water reactor. *Atomic Energy , 47* (5), 916-918.

Erykalov, A. N. (2005). Computed trtium concentration in a heavy-water reactor. *Atomic Energy , 99* (3), 210-214.

"MCNP-A General Monte Carlo N-Particle Transport Code, Vesion 5". LA-UR-03-0245, Los Alamos Nation Labortory (2003).

M. Kalinowski, “International Control of Tritium for Nuclear Nonproliferation and Disarmament”, CRC Press, (2005).

(563b) Continuation of Experts In Discrete Element Modeling (EDEM) Validation Study for Packed Bed Structural Properties

Vaibhav Khane, Chemical Engineering, Missouri University of Science & Technology, Rolla, MO, G E Mueller, Nuclear Engineering Dept., Missouri University of Science & Technology, Rolla, MO and Muthanna H. Al-Dahhan, Chemical Engineering Dept., Missouri University of Science & Technology, Rolla, MO

Experts in Discrete Element Modeling (EDEM)[1] is a discrete element method (DEM) based modeling software program, capable of providing a wealth of information about granular solid flows occurring in many industrial applications. The DEM is a numerical approach used to compute the motion of large number of particles[2]. A computational study involving the application of EDEM to a pebble bed nuclear reactor (PBR) application is being performed at Missouri S&T. Prior to full involvement with EDEM based numerical simulations, an effort has been made to validate the packing algorithm used in EDEM; particularly for cylindrical packed bed applications. This is essential, as the local flow and transport properties are closely coupled with the structural characteristics of a packed bed which in turn depends upon the packing of the particles inside the container. Also, the packing of particles inside the container is the first and main step in EDEM based analysis. Packing algorithms available with commercial codes such as EDEM are used as a 'Black-Box' and are without any detailed validation exercise. In most cases, average porosity results are compared to benchmark the numerical packing results with available experimental results, which is not sufficient. Detailed technical information about the packing algorithm used in EDEM software is not available in open literature and hence there is a need to perform validation studies of the packing algorithm used in EDEM. The packing algorithm used in EDEM demands accurate input of elasticity (material properties) and frictional (interaction properties) parameters which are not readily available in the literature for materials of interest. The radial porosity profile, along with average porosity values, will be used as a means for this validation study. This study also involves a parametric sensitivity study of material and interaction properties on structural properties of cylindrical packed beds. This sensitivity study will highlight important material and interaction properties which need to be provided as an input from a reliable simulation point of view. Also, a DEM based simulation study will optimize the values of material and interaction properties by mimicking an experimentally obtained radial porosity variation profile to a larger extent. Hence, obtained results will provide a new and better understanding about the packing algorithm used in EDEM as well as recommendations to simulate realistic packed bed structures using EDEM. The cylindrical packed bed has been chosen as a model for this validation study. There is much benchmark data available for packed bed structural properties from previous experimental measurements[3] and from high-fidelity numerical packing algorithms4 for a wide range of aspect ratios. Also, the slow and dense granular flow encountered in a PBR can be approximated by static packed beds[5]. The various structural properties of cylindrical packed beds which can be used for this validation study are mean porosity, the distribution of particle centers, and the axially averaged radial porosity profile which exhibits a typical shape of damped oscillation. The radial distribution of particle centers serves as an input to Computational Fluid Dynamics (CFD) analysis of packed beds in order to simulate them more accurately[6]. Radial and axial porosity profiles and associated solid phase distributions are required as an input to hydrodynamic and thermal models of packed beds[7]. It has already been demonstrated in our previous work that the friction between wall and particle plays an important role in the structural characterization of packed beds[8] and neglecting this friction results in simulations of tightly packed bed which are unreal.

DESCRIPTION OF THE ACTUAL WORK

A series of EDEM simulations is being carried out to perform a parametric sensitivity study to check the effect of static friction between particles. The Hertz-Mindlin (with no-slip) contact model[9] has been used to model particle-particle and container wall-particle contact forces. Higher values of static friction coefficient between particles damps out the porosity variation profile quickly while moving towards the center. Neglecting friction between particles under predicts porosity variation near the container wall, resulting in tightly packed beds which are evident from mean porosity values. The effect of other material and interaction properties on the structural properties of packed beds needs to be investigated further. Such study might help in the development of a mechanistic model for the radial porosity variation of packed beds which will require inputs of material properties besides the aspect ratio.

Acknowledgements

The authors acknowledge the financial support provided by Department of Energy (DOE) Nuclear Energy Research Initiative (NERI) project (**NERI-08-043**).

REFERENCES

[1] EDEM™ User Manual, DEM Solutions (USA) Inc., Edinburgh, Scotland (2010).

[2] P.A. CUNDALL, O.D.L.STRACK, A discrete numerical model for granular assemblies. Geotechnique, 29:47–65, 1979.

[3] G. MUELLER, Radial void fraction distributions in randomly packed fixed beds of uniformly sized spheres in cylindrical containers, Powder technology,72,pp.269-275, 1992.

[4] G. MUELLER, Numerically packing spheres in cylinders, Powder technology, 159, pp.105-110, 2005.

[5] C. du TOIT, The Numerical Determination of the Variation on the Porosity of the Pebble-Bed Core, Proceedings of the conference on High-Temperature Reactors, Petten, The Netherlands, European Nuclear Society, 2002.

[6] N.J. MARIANI, W.I. SALVAT, A. CAMPESI, , G.F. BARRETO, O.M. MARTINEZ, Evaluation of structural properties of cylindrical packed beds using numerical simulations and tomographic experiments, International Journal of Chemical Reactor Engineering 7, Art. No. A82.,2009.

[7]A.J. SEDERMAN, P.ALEXANDER and L.F. GLADDEN, Structure of packed beds probed by Magnetic Resonance Imaging, Powder Technology, 117, pp. 255-269, 2001.

[8] V. KHANE, G.E. MUELLER, and M. AL-DAHHAN, Experts in Discrete Element Modeling (EDEM) Validation for Packed Bed Structural Properties, AIChE-2010 Annual Meeting, Salt Lake City, Utah, November 7-12, 2010.

[9] Y. TSUJI, T. TANAKA and T. ISHIDA, Lagrangian numerical simulation of plug flow of cohesion-less particles in a horizontal pipe, Powder Technology, 71 239-250, 1992.

[#] A.G. POLSON, Distinct element method analyses of fuel spheres in the PBMR core using RFC3D, Proceedings of the 2004 International Congress on Advances in Nuclear Power Plants, ICAPP'04, June 13-17, 2004.

See more of this Session: Advances In Numerical Simulations Bridging Chemical and Nuclear Engineering Phenomena or Processes
See more of this Group/Topical: Nuclear Engineering Division

GEANT4 SIMULATION OF IRRADIATION FACILITIES AND NEUTRON SOURCES AT UNIVERSITY OF UTAH TRIGA FOR NUCLEAR FORENSICS AND DETECTION

[1]Andrew Steven Voyles, [2]Haori Yang and [3]Tatjana Jevremovic
Utah Nuclear Engineering Program (UNEP), The University of Utah
[1]Undergraduate student in Nuclear Engineering Minor, [2]Assistant Professor, [3]Chair Professor and UNEP Director
[1]andrew.voyles@utah.edu, [2]haori.yang@utah.edu, [3]tatjana.jevremovic@utah.edu

Introduction

GEANT4 (**GE**ometry **AN**d **T**racking, release 4) high-energy physics simulation toolkit was used to simulate neutron source signatures, as well as various irradiation ports of the University of Utah TRIGA facility (UUTR): central irradiator, thermal irradiator, and some peripheral ports and components. Each of these ports has unique geometry and different neutron and gamma flux. Preliminary results for thermal irradiator and central irradiator, as well as for neutron sources are discussed in this paper, while the modeling of the remaining ports is underway.

Purpose

Being able to quickly and accurately determine the source of various nuclear materials through known nuclear signatures is an essential requirement for successful nuclear forensics. As a part of an ongoing project at UUTR, using GEANT4, we are developing comprehensive nuclear signature data sets and benchmarks against corresponding nuclear experiments. This paper discusses signatures of the various irradiation ports and samples irradiated in such ports at UUTR, signatures for the UUTR fuel element assemblies, and signatures of shielded and unshielded neutron and gamma sources.

Description of the actual work

GEANT4 program flow.

For a general GEANT4 simulation, a few files are adapted to fit the end user's run-time parameters, namely, the geometry of the detector and solids involved in the simulation, how particles are generated, and with what physical properties, and how the particles will propagate through space. At the time of compilation, these header files are read into the execution script, and complied to produce the executable of the simulation.

Once executed, GEANT4 provides two routes for program flow: interactive and batch modes. For batch mode, a macro file, written ahead of time, is read into the idling command prompt in the simulation, and executed lien-by-line. This is convenient for simulations that are compiled ahead of time, and simply running the macro file will carry out the entire simulation autonomously, without any further input. Another alternative is interactive mode. In this mode, the user can execute commands into the idling command prompt to achieve the exact same results as batch mode, for a macro containing the same commands. This mode is useful for debugging a simulation, or for just experimenting. However, both modes can be used to attain the same desired results. Once the simulation is completed, the user can then use the commands of either OpenGL or HepRep to output the visualization of their simulation. The simulation is designed to be general for all common simulation applications. End users will only need to supply data on the geometry and composition of their setup and samples in the *GeometryConstruction.cc* file, as well as the neutron flux values at runtime, and the simulation will be carried out in batch mode, and visualized.

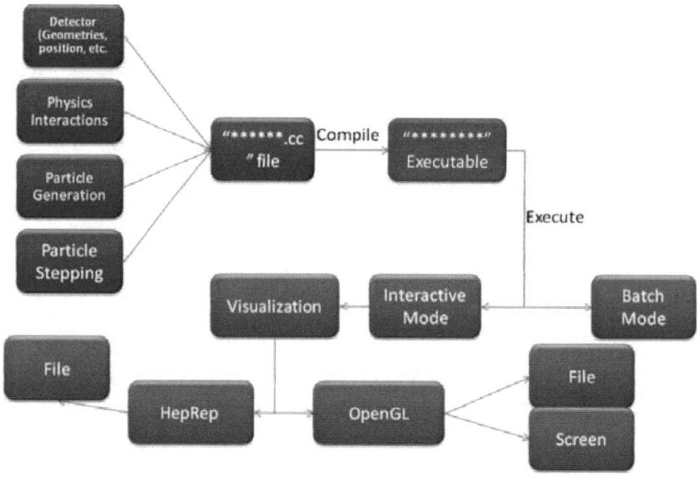

Figure 1. Program flow for generalized GEANT4 simulation.

GEANT4 detection and scoring.

Calculations of absorbed dose at each distance from the source were accomplished via use of multiple sensitive detectors in GEANT4. For example, 5cm thick germanium detectors were constructed at 25, 50, 75, 100, 125, 150, 175, and 200 cm away from the center of the sample source. These detectors were assigned logical scorers in GEANT4, made sensitive only to gamma photons. At the end of each simulation, the detectors calculate the accumulated gamma dose in each detector, and these results are printed to the command line, as well as to a file, designed for easy importing of data for later analysis and visualization. An example of the detector array, constructed around the UUTR central irradiator can be seen in Figure 2.

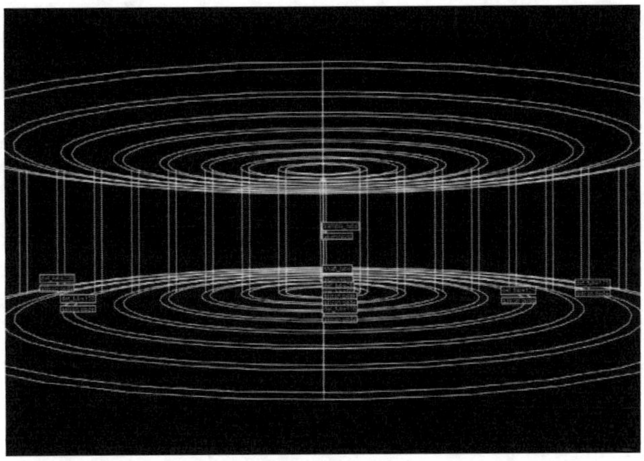

Figure 2. Example of detector array, constructed around central irradiator model. The same array of detectors was used for both the thermal and central irradiator models. Central model used here for illustration purposes.

GEANT4 model of neutron energy spectra.

The Watt model of fission neutron energies was used to model the thermal neutron flux in the various port models. This model is described in Equation 1, and a plot of the fraction of fission neutrons per MeV as a function of the neutron energy is seen in Figure 3 below.

$$\chi(E) = 0.453e^{-E}\sinh{(2E)}^{0.5}$$

Equation 1. Watt Model of Fission Neutron Energy distributions.

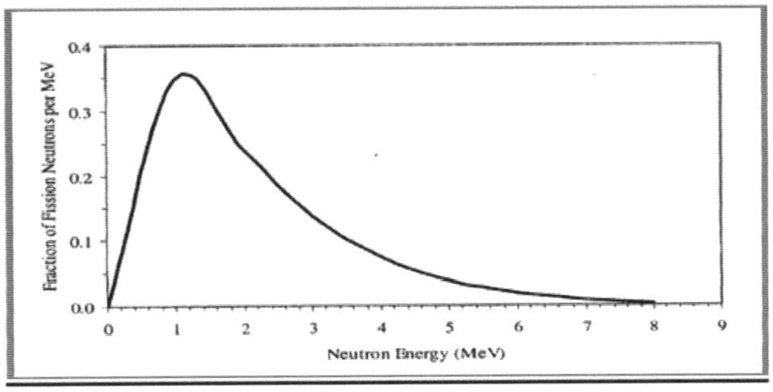

Figure 3. Watt Fission Neutron Energy distribution. Image taken from Jevremovic, "Nuclear Principles in Engineering" [4].

GEANT4 model of UUTR central irradiation port.

Figure 4 shows the model of a 1"-diameter, 5"-tall aluminum cylinder in the center of a 15.24 cm-diameter, 38.1 cm-tall cylindrical aluminum air shaft, subjected to a flux of 3.6 $*10^{12}$ $\text{neutrons}/_{\text{cm}^2 * \text{s}}$. MCNP5 was used to simulate this same geometry, to obtain neutron flux-per-source-particle, to calculate the number of neutrons necessary to be generated in order to simulate a desired flux in GEANT4.

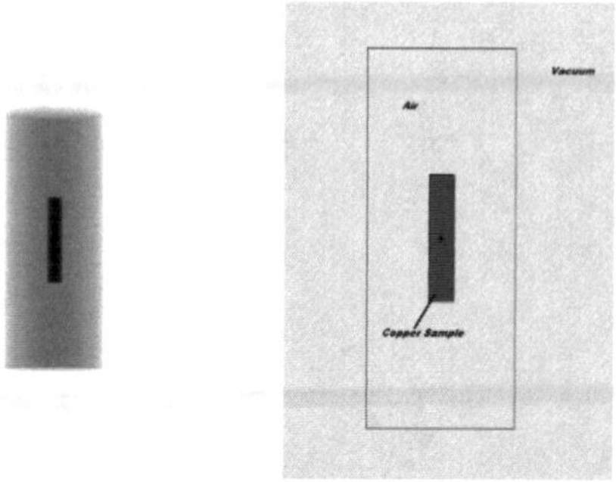

Figure 4. MCNP5 Model of a sample placed in central irradiator (CI); flux data used to run the corresponding GEANT4 model

GEANT4 model of UUTR thermal irradiation port.

Figure 5 shows the model of a 1″-diameter, 5″-tall aluminum cylinder 1″-diameter, 5″-tall aluminum cylinder in center of a 1″-diameter, 35.08-cm tall aluminum air shaft. Air shaft nested at the center of 11.4-cm diameter, 35.08*cm tall shaft of heavy water. Shafts embedded in aluminum trapezoidal prism with bases of 29.8cm and 47.4cm, 15.2cm wide, and 60cm tall, filled with heavy water. Aluminum volumes are 1/16″ thick. Sample was subjected to a flux of $3.6 * 10^{11}\ \mathrm{neutrons}/_{\mathrm{cm}^2 * \mathrm{s}}$.

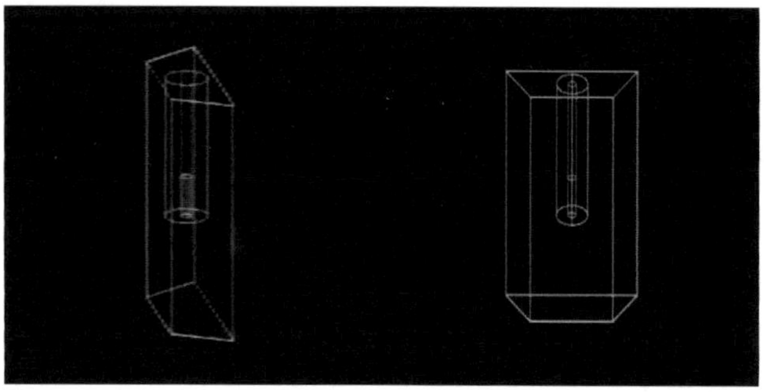

Figure 5. GEANT4 Model of an aluminum sample placed in thermal irradiator (TI)

GEANT4-based UUTR nuclear signatures.

Calculations of average neutron and gamma photon energy spectra were accomplished via tracking neutrons and gamma photons in desired volumes, specific to each of the above presented models. For the thermal irradiator and central irradiator models, the average neutron energy spectra were calculated from the neutrons in the air shafts leading up from the irradiation ports, and the average gamma photon energy spectra were calculated from the gamma photons in the samples in the irradiation ports. For the shielded and unshielded Pu-Be neutron source models, the average neutron energy spectra and the average gamma photon energy spectra were calculated from the neutrons and gamma photons, respectively, penetrating the inner surface of the inner, gamma photon detector.

Results

GEANT4 visualizations of UUTR central and thermal irradiation ports.

In the GEANT4 simulation model, the aluminum cylinder was exposed to 3.6 $*10^{12}$ $\mathrm{neutrons}/_{\mathrm{cm}^2 * \mathrm{s}}$, homogeneous cylindrical flux, using 1.07745 $*10^{-3}$ $\mathrm{neutrons}/_{\mathrm{cm}^2 * \mathrm{s}}$ per neutron. Figure 6 illustrates the homogeneous neutron flux for an irradiated aluminum sample, simulated in GEANT4.

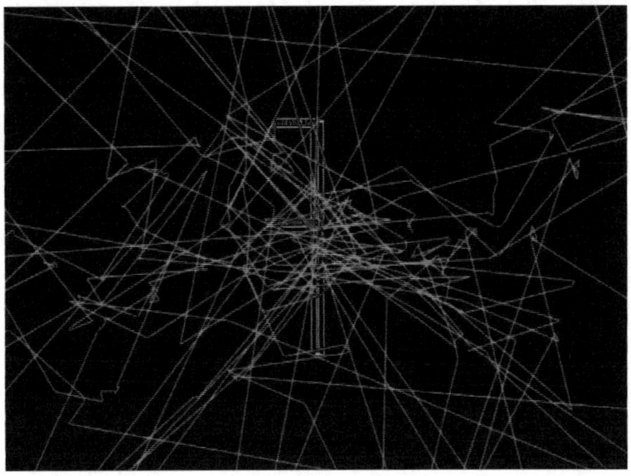

Figure 6. GEANT4 simulation of central irradiator irradiation of aluminum sample. Green rays indicate neutron and gamma tracks; red tracks indicate negatively-charged particles; blue tracks indicate positively-charged particles; white points mark stepping points for particles.

In the GEANT4 simulation model, the aluminum cylinder was irradiated with 3.6 $*10^{11}$ $\mathrm{neutrons}/_{\mathrm{cm}^2 * \mathrm{s}}$, homogeneous planar flux. Figure 7 illustrates the homogeneous neutron flux for an irradiated aluminum sample, simulated in GEANT4.

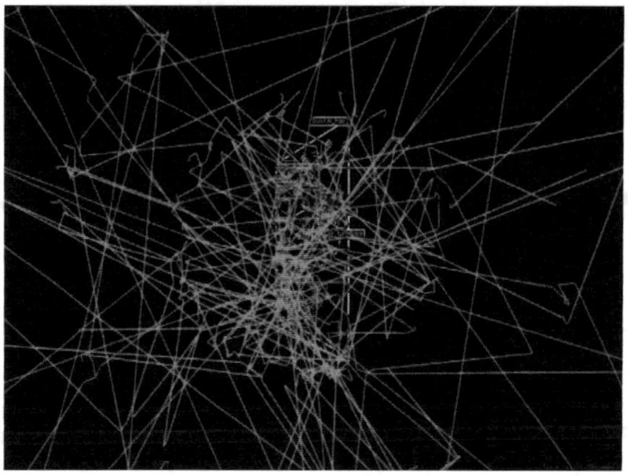

Figure 7. GEANT4 simulation of thermal irradiator irradiation of aluminum sample. Green rays indicate neutron and gamma tracks; red tracks indicate negatively-charged particles; blue tracks indicate positively-charged particles; white points mark stepping points for particles.

GEANT4 dose calculations for UUTR central and thermal irradiation ports.

In GEANT4, a particle flux is simulated by generating various large (>>10,000) numbers of source particles. The dependent variable, in this case, the dose rate at various distances from the irradiated sample sources, is then calculated for each run of particles. If a linear or some other relationship can be drawn between the number of source particles generated and the dependent variable, it is reasonable to be able to scale up the results of the dependent variable with respect to the number of source particles generated. In this case, the absorbed dose rate varied linearly with respect to the number of source particles for both the CI and TI. Thus, many runs of particles were simulated at various numbers of source particles between 100 and 1,000,000 source neutrons. This data was plotted to generate a best-fit regression line between the numbers of source particles and the absorbed dose rate. To compare results, the best-fit dose for 100,000 neutrons was used in the results for both the CI and TI models, discussed below. The dose vs. numbers of source particles plots for the TI model can be seen in Figure 8. From Figure 8, the mean dose for 100,000 neutrons generated was calculated for each distance from the source. This dose rate result is then scaled up by factor of $\frac{\text{desired actual flux}}{\text{flux for 100,000 neutrons}}$, to give results comparable to what should be seen experimentally. These absorbed dose rates were tabulated at each distance to yield a surface plot of dose rate as a function of distance from the source, seen in Figures 9 and 10.

As an example calculation, at 25cm from the irradiated sample, the absorbed dose for 0.154 μs irradiation time of aluminum sample from thermal irradiator is $5.995 \cdot 10^{-15}$ Gray. For a 2 hour irradiation time common for NAA, this equals a dose rate of $1.39 \cdot 10^{4}$ $^{mrem}/_{hr}$. For comparison, at University of Utah TRIGA, Area Radiation Monitors SCRAM reactor for detection of > 10 $^{mrem}/_{hr}$ dose rates. In this case, samples are suspended at half depth in reactor pool to cool until < 10 $^{mrem}/_{hr}$. For 1 $^{mrem}/_{hr}$ < dose rates < 10 $^{mrem}/_{hr}$, samples are cooled inside 4″-thick lead box until < 1 $^{mrem}/_{hr}$. Samples < 1 $^{mrem}/_{hr}$ are deemed safe. The central irradiator model displays the similar results, for a cylindrical neutron source.

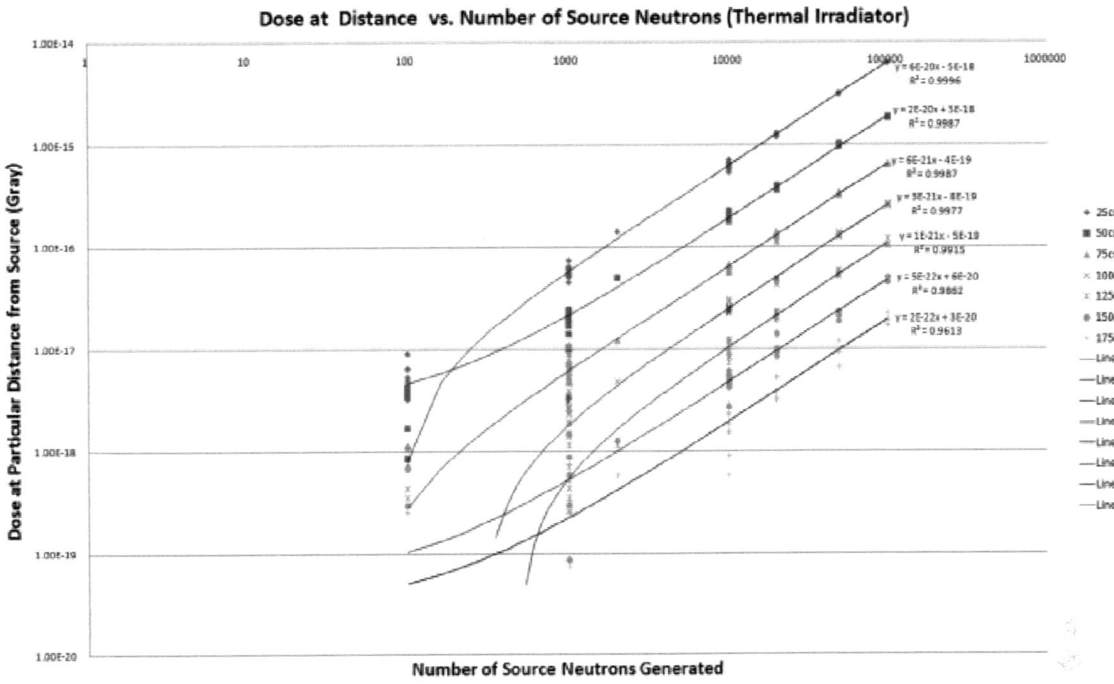

Figure 8. GEANT4 dose at various distances from sample as a function of the number of source particles generated for the Thermal Irradiator model. At each distance, for > 100 neutrons generated, there is a strong linear relationship between the independent and dependent variables.

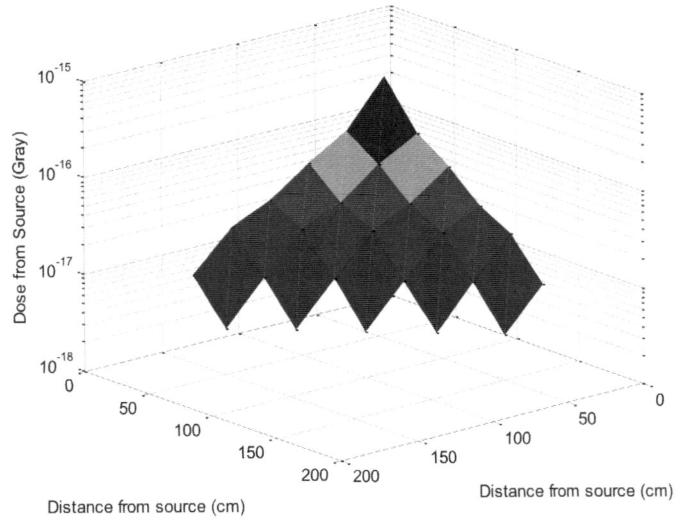

Figure 9. Dose plot for the thermal TRIGA irradiator model, as a function of distance from the source, corresponding to a 0.154 µs irradiation time. Image taken from Voyles et al. [7].

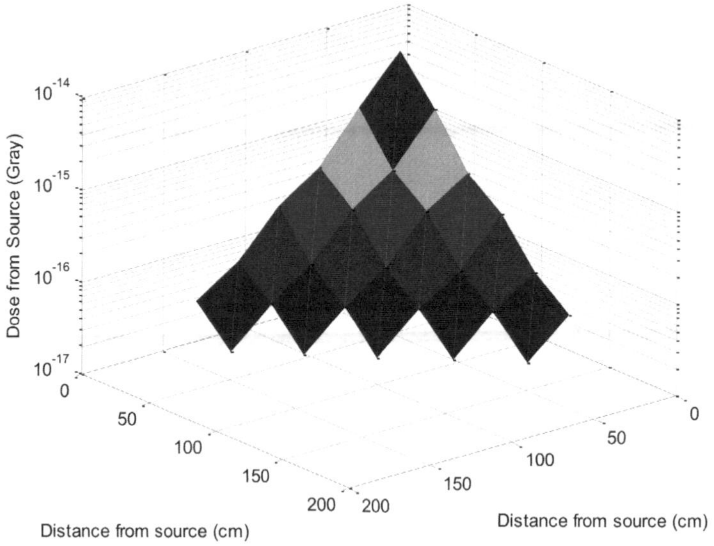

Figure 10. Dose plot for the central UUTR irradiator model, as a function of distance from the source, corresponding to a 0.0141 μs irradiation time. Image taken from Voyles et al. [7].

GEANT4 neutron source simulations.

GEANT4 simulations of an unshielded Pu-Be neutron source, as well as simulations of the same source, shielded with paraffin and lead, have been created. Figure 12 shows that very few of the green (neutrally-charged) tracks make it outside the paraffin shell. These tracks represent the neutrons and gamma photons, and so it is clear that the paraffin is a good moderating material for the neutrons (and slightly so for the gamma photons as well).

GEANT4 neutron and gamma signatures.

Average neutron and gamma photon energy spectra were created for the thermal irradiator, central irradiator, shielded Pu-Be neutron source, and unshielded Pu-Be neutron source models, using the methods described above, in "GEANT4-based UUTR nuclear signatures". After tracking gamma photons and neutrons being transported in the volumes specific to each model (described in the aforementioned earlier section), the kinetic energies of these particles were stored, and compiled into histograms for each of the thermal irradiator, central irradiator, shielded Pu-Be neutron source, and unshielded Pu-Be neutron source models. These histograms are shown in Figures 12, 13, 14, and 15.

For the central irradiator, as well as the shielded and unshielded Pu-Be neutron source models, gamma photon and neutron energy spectra for various simulation models, no experimental data were available for benchmarking. However, it is very obvious that in most of the models, excessive thermalization of neutrons, as well as somewhat for gamma photons, generated very indistinct gamma photon and neutron energy spectra. Preliminary benchmarking has been accomplished, using a 150-mg sample of limestone ($CaCO_3$), irradiated in the thermal UUTR irradiator for 2 hours, and then allowed to decay for 72 hours.

Figure 16 shoiws the experimental gamma photon energy spectrum, as well as the gamma photon energy spectrum for the simulation of this irradiation.

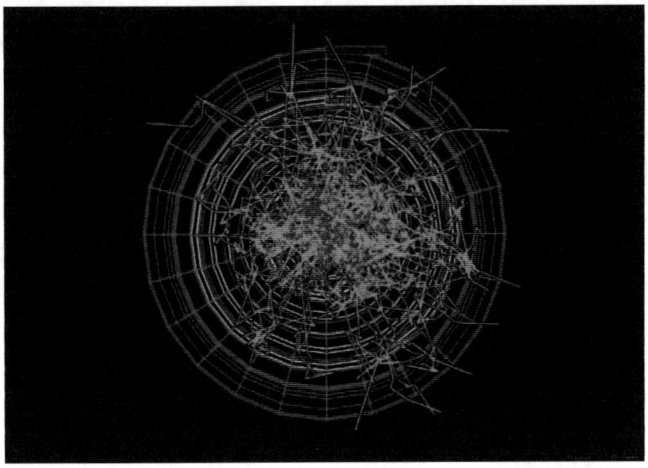

Figure 11: Top-down view of experimental setup for 6.700 · $10^6 \, \mathrm{neutrons}/_{\mathrm{cm}^2 * \mathrm{s}}$, with particle tracks. Red lines indicate negatively-charged particle trajectory, green lines indicate neutrally-charged particle trajectory, blue lines indicate positively-charged particle trajectory, yellow dots indicate the actual step points used by Geant4.

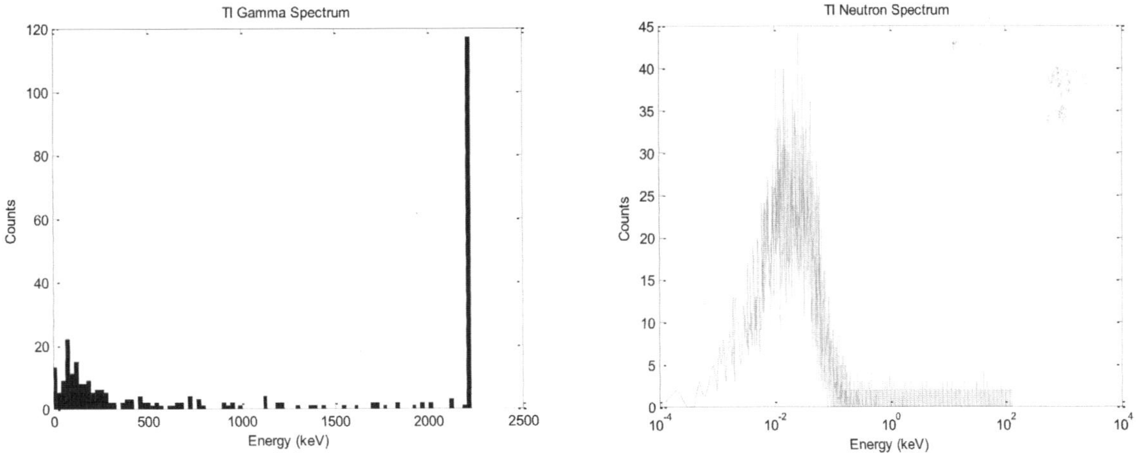

Figure 12. Average gamma photon and neutron energy spectra for the thermal TRIGA irradiator model of irradiation of aluminum sample, seen n Figure 7.

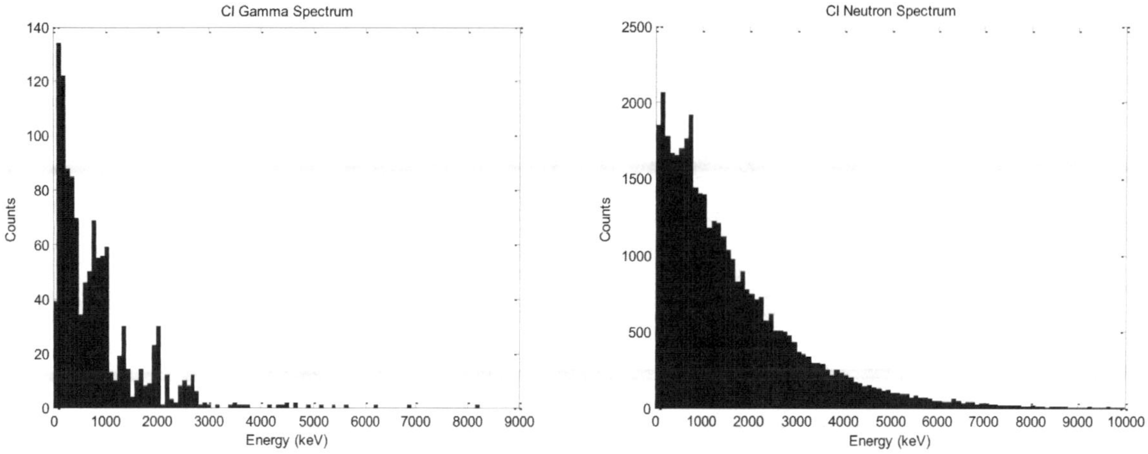

Figure 13. Average gamma photon and neutron energy spectra for the central TRIGA irradiator model of irradiation of aluminum sample, seen in Figure 6.

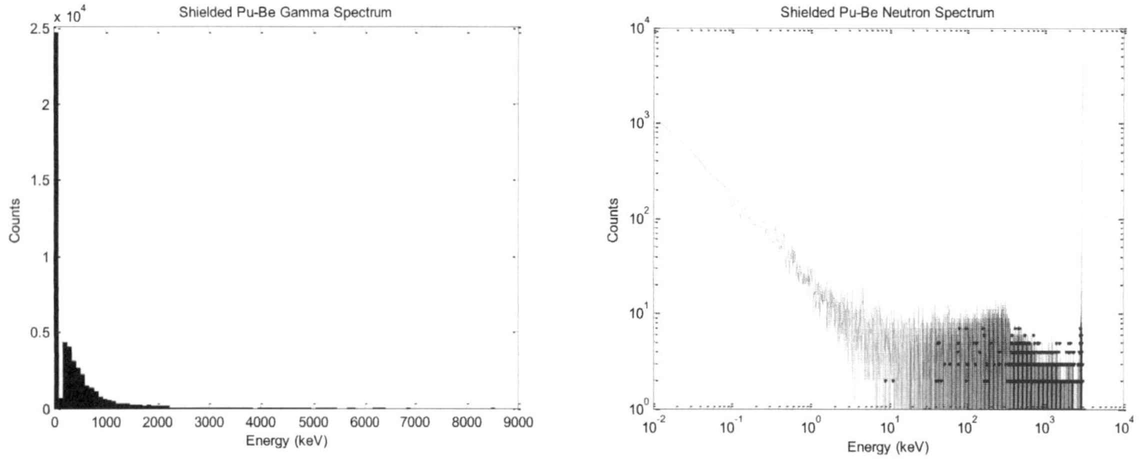

Figure 14. Average gamma photon and neutron energy spectra for the shielded Pu-Be neutron source model of irradiation of aluminum sample, seen in Figure 12.

As can be seen, there is some basic correlation between the experimental and simulation spectra, in Figure 17. The largest discrepancy comes in the large peak at ~2200 keV. This most likely comes from the presence of hydrogen nuclei in the simulation, and can be seen distinctly in the gamma photon spectrum for the thermal irradiator model, seen in Figure 13. This issue with such neutron activation experiments has been documented in literature, by Evans et al. [1]. Future work in this simulation will address and correct this issue.

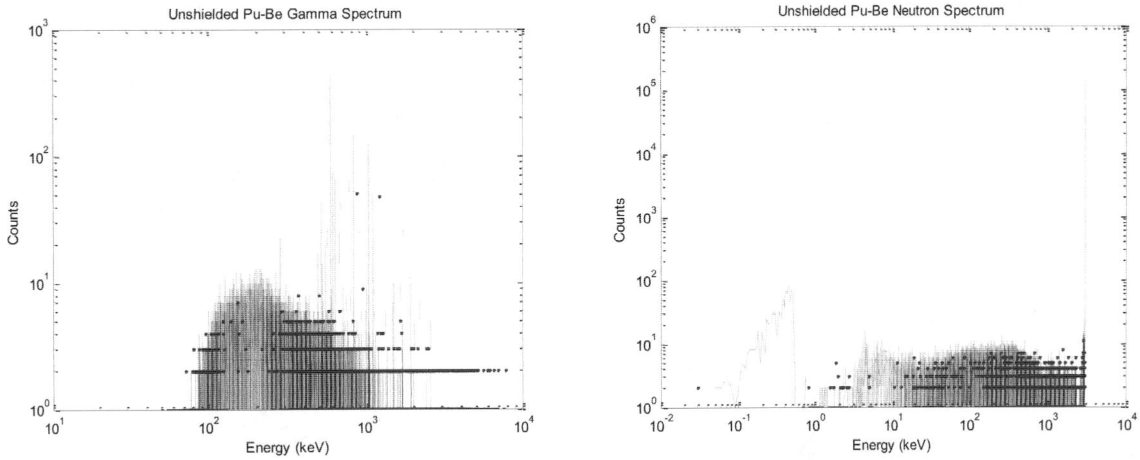

Figure 15. Average gamma photon and neutron energy spectra for the unshielded Pu-Be neutron source model of irradiation of aluminum sample.

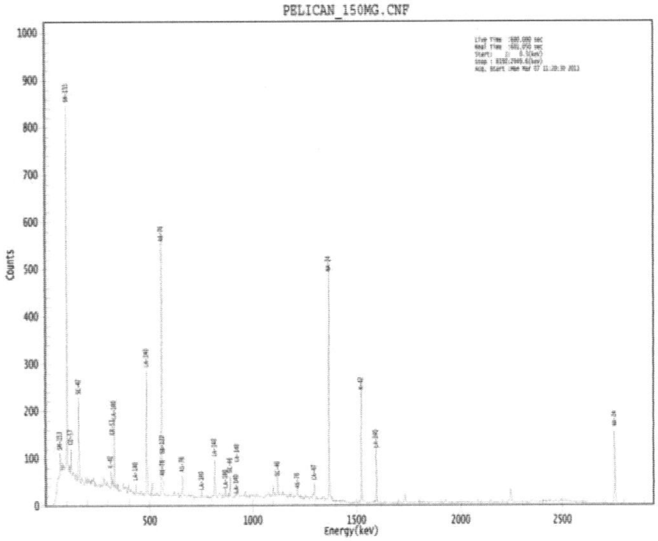

Figure 16: Experimental gamma photon energy spectrum for 150-mg limestone sample irradiated in UUTR TRIGA thermal irradiator for 2 hours.

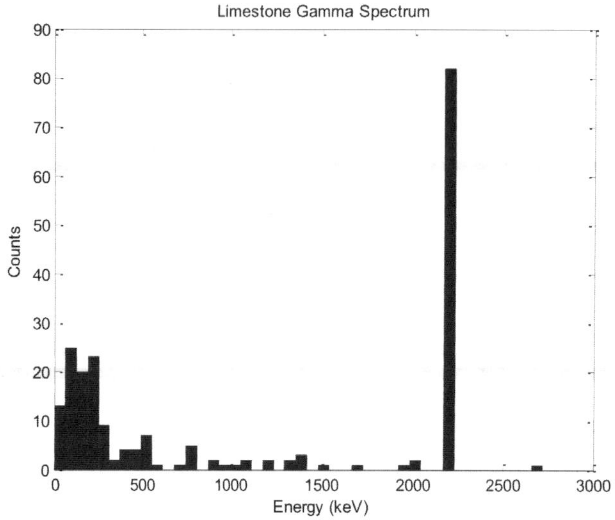

Figure 17: Experimental gamma photon energy spectrum for 150-mg limestone sample irradiated in UUTR TRIGA thermal irradiator for 2 hours.

Conclusion

A preliminary GEANT4 model (coupled with MCNP5 simulation for accurate representation of the neutron flux) was developed for a central irradiation port and thermal irradiation port at Utah TRIGA facility, as well as shielded Pu-Be neutron and gamma sources. Using MCNP5 to calculate flux-per-particle, our current GEANT4 model can simulate the irradiation of various samples. The next step is to expand the current model to include the fast neutron irradiation facility at Utah TRIGA facility. In addition, simulation results will be benchmarked against experimental results, to help make this application suitable for export, in the use of helping benchmark experiments in new facilities.

Acknowledgement

This research is partially funded by the DHS/DOE Nuclear Forensics Education Award Program.

References

1. Evans, C. J., Ryde, S. J. S., Hancock, D. A., and Al-Agel, F. "Monte Carlo Simulation of Prompt Gamma Neutron Activation Analysis Using MCNP Code". Applied Radiation and Isotopes, Vol. 49, No 5/6, pp. 541-543, 1998.

2. Geant4: a toolkit for the simulation of the passage of particles through matter. *CERN* - the European Organization for Nuclear Research, 2011. http://www.geant4.org/geant4/
3. HepRep: a Generic Interface Definition for HEP Event Display Representables. Joseph Perl, Stanford Linear Accelerator Center, 2009. http://www.slac.stanford.edu/~perl/heprep/index.html
4. Jevremovic, Tatjana. "Nuclear Principles in Engineering." New York: Springer Science+Business Media, Inc., 2008.
5. MCNP - A General Monte Carlo N-Particle Transport Code - Version 5. Monte Carlo Codes Group, Los Alamos National Laboratory, 2006. http://mcnp-green.lanl.gov/
6. OpenGL - The Industry's Foundation for High Performance Graphics. Khronos Group, 2010. http://www.opengl.org/
7. Voyles, Andrew S., Yang, Haori, and Jevremovic, Tatjana. "GEANT4 Simulation of Irradiation Facilities at University of Utah TRIGA." Unpublished paper presented at American Nuclear Society Student Conference, Atlanta, GA, 2011.

(563e) Force Field Parameterization and Monte Carlo Simulation of Alkoxy-Ethanol Surfactant

Sunghyun Jang, Junhyuk Lim and Hwayong Kim, Department of Chemical and Biological Engineering, Seoul National University, Seoul, South Korea

Alkoxy-ethanolis highly associating molecule which contains ether (-O-) and hydroxyl (-OH)group in the same molecule. It has both inter- and intra-molecular hydrogenbonding, thus the complex interaction makes it difficult to estimate thethermophysical properties of its fluid mixtures. When simulating the pure alkoxy-ethanolor its mixture system, the positive charge of hydroxyl hydrogen plays animportant role because it could form partial hydrogen bonding with both theproton accepters.

The TraPPE and NERDforce filed is the two most widely used force fields which provide the robustmolecular simulation framework for various thermophysical properties'estimation, especially phase equilibrium study. These two force fields use theunited atom based approach to represent the CH, CH2, CH3 groups which enables thetransferability of their force field parameters to a wide range of molecularspecies. With this simulation methodology, coexistence properties can bepredicted with relatively high levels of accuracy. Basically these two forcefields have very similar energy expression each other, but they don't considerthe hydrogen bonding term explicitly in its energy expression. Therefore theeffect of hydrogen bonding should be investigated for some specific system suchas alkoxy-ethanol and its mixture system.

On this unitedatom base, the force field parameters of alkoxy-ethanol have been optimized andthe molecular structure and coexisting properties of alkoxy-ethanol and itsmixture has been studied. The results are compared to simulation results usingNERD and TraPPE parameters.

See more of this Session: Advances In Numerical Simulations Bridging Chemical and Nuclear Engineering Phenomena or Processes
See more of this Group/Topical: Nuclear Engineering Division

GEANT4 MODEL FOR NUCLEAR FORENSIC ANALYSIS OF NUCLEAR DETONATION IN URBAN ENVIRONMENTS

Jeffrey Taylor, Haori Yang, Xuesong Zhou, and Tatjana Jevremovic
University of Utah, Salt Lake City, UT, USA

Introduction

Nuclear forensics techniques are commonly applied in pre-detonation circumstances to help identify the material characteristics and source of intercepted material. Just as important, the field of post-detonation forensics applies similar techniques for device attribution. This paper summarizes research and work relevant to creating a simple model in GEANT4 for post-detonation nuclear forensics applications in urban environments. The simulation model incorporates an intense neutron and gamma ray pulse source located in an urban canyon landscape in downtown Salt Lake City with approximate building dimensions to simulate a low-yield nuclear device detonation. Following a quick review of nuclear forensics in urban environments, this paper characterizes model inputs, such as prevalent material types and the radiation source, describes the modeling approach in GEANT4, and discusses simulation results.

Nuclear Forensics in Urban Settings

A report published by the American Physical Society and the American Association for the Advancement of Science (AAAS) defines nuclear forensics as "... the technical means by which nuclear materials, whether intercepted intact or retrieved from post-explosion debris, are characterized (as to composition, physical condition, age, provenance, history) and interpreted (as to provenance, industrial history, and implications for nuclear device design)" (9). The primary objective of nuclear forensic analysis is to identify the material characteristics to help determine the material's origin, intended use, and the method of acquisition, with the ultimate goal being attribution. Since government agencies often focus their search for illicit nuclear material away from major urban areas (ports, borders, etc.), nuclear forensics in urban environments is often more focused toward post-detonation attribution.

In an urban detonation scenario, it may be assumed that a low-yield fissile device detonates in an urban canyon environment, with many large buildings in the immediate vicinity. Depending on the size and type of the device, there may be sufficient residual material for forensic analysis, but there are situations which may preclude this option (i.e. small material source experiencing significant dispersion after detonation). In these situations, attempting to identify the source material may rely upon measuring the prompt gamma rays and neutrons, which are released within the first seconds after the detonation, and comparing the observed spectral distributions with simulation results for a similar environment. This approach requires simulating the source and the surrounding environment, which obviously increases the uncertainty associated with the resulting source characterization estimates compared to physical forensic analysis.

Source Characterization Simulation

Simulation tools used for source characterization typically face a significant challenge modeling the source because they are described as a device. Modeling the source as a device requires consideration for the geometry, material construction, method of detonation, and other characteristics of the weapon, all of which are unknown and significantly impact the simulation results. Additionally, this approach requires modeling many potential combinations of device characteristics, increasing analysis time and further complicating the simulation process.

Since the source is the largest unknown element, the best option for limiting the remaining sources of additional uncertainty requires creating a more realistic simulated environment. Interactions with structures and other materials in the built environment affect the observed radiation measurements, so these effects should be considered in the simulation. As a result, creating this type of simulation requires knowledge of the built environment near the detonation site, including the dimensions and materials used in construction. This level of detail, which is greater than what is typically available for emergency response and analysis simulations, is particularly well-suited for implementation with a Geographic Information System (GIS). A GIS is a system for managing and analyzing spatially-referenced data which can associate attributes with objects with a defined shape and geographic location.

Modeling Approach

To address the issues associated with source characterization simulations described above, the proposed modeling approach in this paper simplifies the source model while utilizing a higher-detail environment model. The source material is simplified to a single point source with a user-defined spectral energy distribution and time-dependent intensity. This reduces modeling complexity by bypassing most device characteristics, reducing the size of the potential solution set (at the cost of increasing uncertainty). The surrounding environment is constructed from input data stored in a GIS, containing building geometries and object attributes (i.e. building with dimensions, shape, material, etc.). This provides the capability to model highly detailed urban areas. Lastly, the simulation is constructed within GEANT4 so that the simulation can fully incorporate all environment characteristics, such as particle interactions and shielding effects, which may affect measurement results. The remaining portion of this paper describes the process of building a simple model for simulation in GEANT4.

Model Inputs

As described above, the primary inputs for this simple GEANT4 model can be divided into the source characteristics and built environment characteristics. The source characteristics are defined by the time-based intensity of the source and the spectral energy distribution for prompt gamma rays and neutrons from a fission event. The built environment characteristics can be defined primarily by the structures in the urban environment, where each building has a center location, dimensions, and an assigned material composition.

Gamma Ray and Neutron Source Characterization

Research related to nuclear weapons testing in the 1950s was referenced to estimate the spectral energy distributions and time-based intensity of prompt gamma rays and neutrons from a small detonation. As seen in Figure 1, a peak in gamma rays occurs almost immediately after detonation (within a microsecond), followed by a decreasing trend in intensity. Fission products decaying after detonation, as well as neutron interactions with the atmosphere, contribute to more gamma rays being emitted after the initial prompt gamma rays when compared to a high altitude detonation (dashed line in Figure 1). Glasstone and Dolan describe a similar peak in neutrons emitted after detonation, where "… 99% of fission neutrons are produced almost immediately, within less than a millionth of a second of the initiation of the explosion" (340). From these observations, modeling the source as an intense pulse from a point source is a reasonable approximation for this model.

Characterizing the source spectral energy distribution for prompt gamma rays and neutrons, Glasstone and Dolan also provide very rough energy distribution histograms, as seen in Figure 2 and Figure 3. The gamma ray spectral distribution in Figure 2 shows nearly 70% below 750 keV, producing a large peak at the low side of the energy spectrum, with a relatively flat distribution at the remaining higher energies. The neutron spectral distribution in Figure 3 shows a relatively flat distribution at energies below 1 MeV, followed by a trend generally showing a nearly-linear decrease in the distribution at higher energy levels until a sharp decrease at 10 MeV. These spectral distributions were identified as examples in Glasstone and Dolan because each device is slightly different, so these distributions should not be considered characteristic of all detonations from fission events. In this case, they are only used to inform the creation of the initial energy spectrum for the model.

Urban Building Material Characterization

Materials in the urban environment vary from the steel and concrete buildings to the trees, cars, and people passing through the city. Since a static geometry model was used in this GEANT4 model, attention was drawn particularly to buildings, since they are expected to act like shielding materials within the environment and thus greatly influence the simulation results. Typical building materials in modern cities often include combinations of concrete, masonry, steel, wood, and glass. This model focused primarily on the major structural components of most buildings, specifically concrete, stone, brick, and steel. These four materials are composite materials, composed of several compounds with various types and compositions depending on the site of their creation, their purpose, and the personnel mixing the materials, requiring definition in the GEANT4 DetectorConstructor source file. Several composite materials are pre-defined in the GEANT4 materials database, but of the four materials defined in the model, only concrete was found in the database. The remaining three materials required definition.

Defining a composite material in GEANT4 requires the fractional mass (or percent composition) of the individual elements or compounds found in the composite material and the overall density of the material. Table 1 describes these general characteristics for each of the four materials. Since there are multiple types of stone, brick, and steel, more common building materials such as granite, red brick and A36 structural steel were chosen for their use in the model.

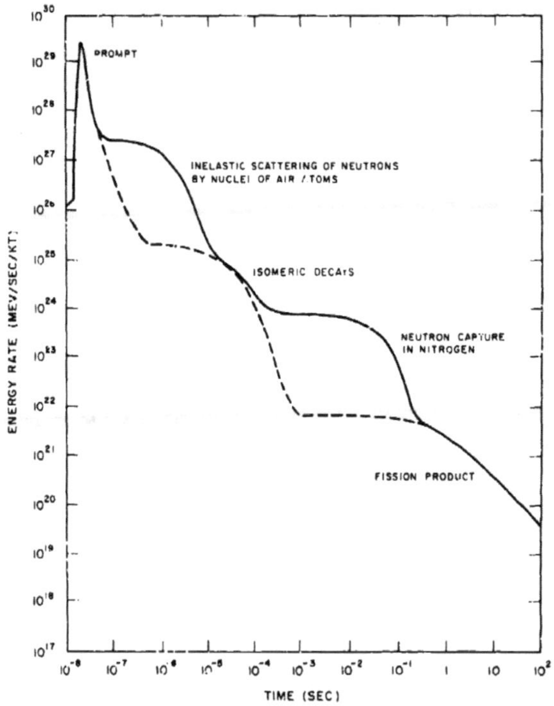

Figure 1. Temporal gamma ray intensity distribution after hypothetical detonation (Glasstone and Dolan 328)

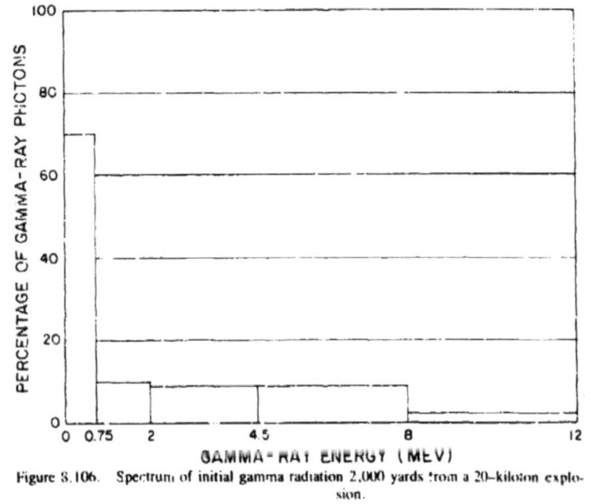

Figure 2. Example of initial gamma ray spectral energy distribution (Glasstone and Dolan 361)

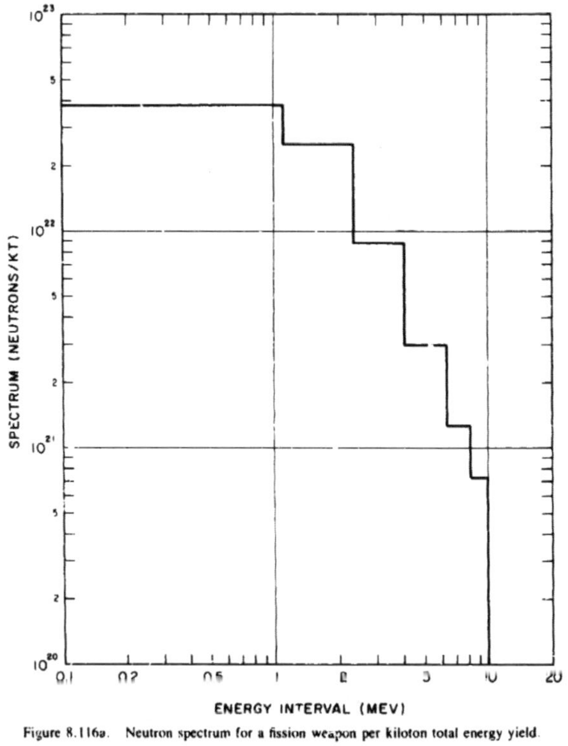

Figure 3. Example of initial neutron spectral energy distribution (Glasstone and Dolan 364)

Table 1. Fractional Mass Composition & Densities for Concrete, Stone, Brick, and Steel

Concrete[1]		**Stone (Granite)**[2]		**Red Brick**[3]		**Steel (A36 Stuct.)**[4]	
Element	% Comp.	Element	% Comp.	Element	% Comp.	Element	% Comp.
O	0.529	O	0.484	O	0.491	Fe	0.982
Si	0.337	Si	0.337	Si	0.280	Mn	0.010
Ca	0.044	Al	0.076	Al	0.159	C	0.003
Al	0.034	K	0.034	Fe	0.035	Si	0.003
Na	0.016	Na	0.027	Ca	0.032	Cu	0.002
Fe	0.014	Fe	0.022	Mg	0.003	Density = 7.85 g/cm^3	
K	0.013	Ca	0.013	Density = 2.0 g/cm^3			
H	0.010	Mg	0.007				
Mg	0.002	Density = 2.7 g/cm^3					
C	0.001						
Density = 2.3 g/cm^3							
Sources: [1] Geant4 User's Guide for Application Developers; [2] Blatt and Tracy; [3] Punmia et al.; [4] MatWeb							

Building Geometries and Locations

Most of the building locations and their dimensions were acquired by taking measurements in ESRI's ArcGIS using aerial photography available from the GIS Portal of the Utah Automated Geographic Reference Center. After determining the world environment's center location in the X-Y plane, building height information was acquired from the Emporis website, featuring lists of the tallest buildings in major cities with their heights. This manual process of connecting a building with attribute data in a GIS is visualized in Figure 4, where the green blocks represent the buildings with height data available, and the shape dimensions were defined by the spatially- and size-accurate aerial photography. The green blocks with light blue highlighting identify the buildings added to the GEANT4 DetectorConstruction file for this analysis. The building position can be estimated by measuring between the world and shape centers. The end result is a table of half-width sizes in the X, Y, and Z directions, as well as vector positions for the center of each building in reference to the world center, as seen in Table 2.

Predominant Physics in Urban Environments

The GEANT4 simulation is modified to consider particle interactions through attenuation, scattering, and absorption for the composite materials in the model. Using data available from the National Institute of Standards and Testing (NIST) and the Korean Atomic Energy Research Institute (KAERI), attenuation coefficients and cross sections were plotted for

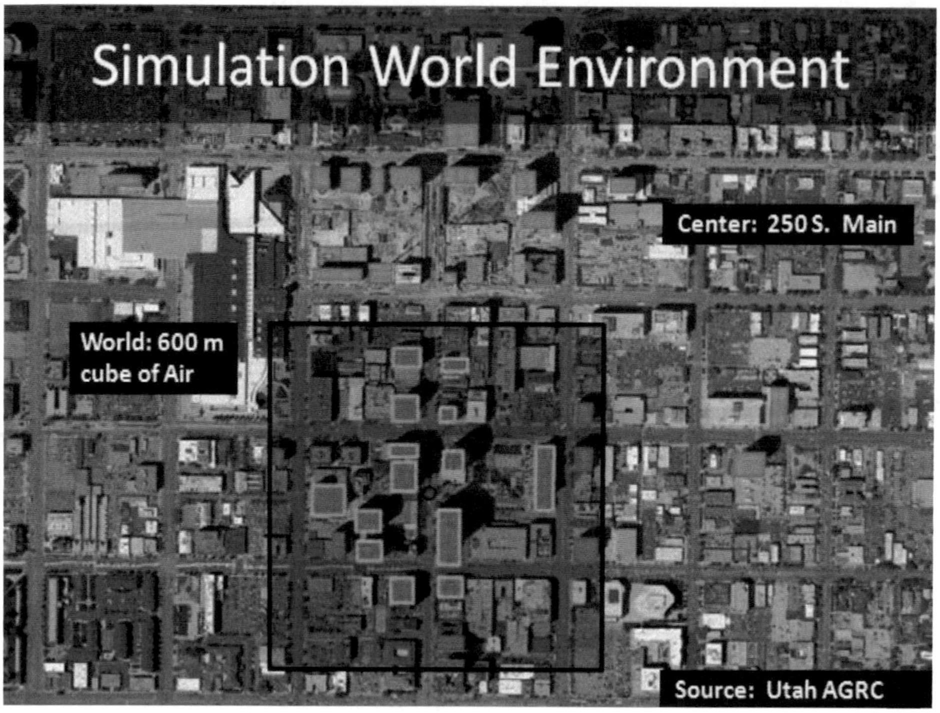

Figure 4. Small area in Salt Lake City as the basis for the simulation world environment

Table 2. GEANT4 Model Building Geometry Characteristics

Name	Half_Width (X)	Half_Length (Y)	Half_Height (Z)	X_Center	Y_Center	Z_Center
Wells Fargo	20	50	65	35	-75	65
One Utah Center	20	18	54	45	58	54
American Tower North	23	18	50	-113	-48	50
American Tower South	23	18	50	-108	-103	50
Walker Center	15	13	34	30	138	34
Hotel Monaco	25	10	33	-50	75	33
Tribune Bldg.	23	10	22	38	225	22
Judge Bldg.	20	18	16	35	-163	16
US Bank	23	23	23	-48	148	23
Kearns Bldg.	25	15	22	-45	240	22
222 Main	23	25	48	-48	30	48
Marriott Center	18	58	19	208	33	19
Hilton Center	30	25	28	-185	-10	28
JC Penney	23	23	23	-48	-168	23

gamma rays and neutrons, respectively, across the energy spectrum. These plots help to indicate which interactions are most likely to take place when gamma rays and neutrons with various energies come in contact with various materials in the urban environment.

Gamma Ray Interaction

Gamma attenuation coefficients from the NIST for concrete and steel are plotted in Figures 5 and 6, where the plots identify different coefficients for total attenuation, coherent and incoherent scattering, photoelectric absorption, and pair production between 1 keV and 20 MeV. Since the composition of brick, stone, and concrete are quite similar, and upon inspection were nearly identical, concrete was used here to visualize the attenuation coefficients representative of all three materials. As seen in Figure 5 below for concrete, the photoelectric effect is expected to dominate below 30 keV and incoherent scattering should be dominant between 30 keV and approximately 20 MeV, with pair production taking precedence at energies above 20 MeV. In Figure 6, the same interactions are expected to dominate in nearly the same energy levels, with the upper bound for photoelectric absorption slightly higher than for concrete, and pair production's lower bound is closer to 10 MeV.

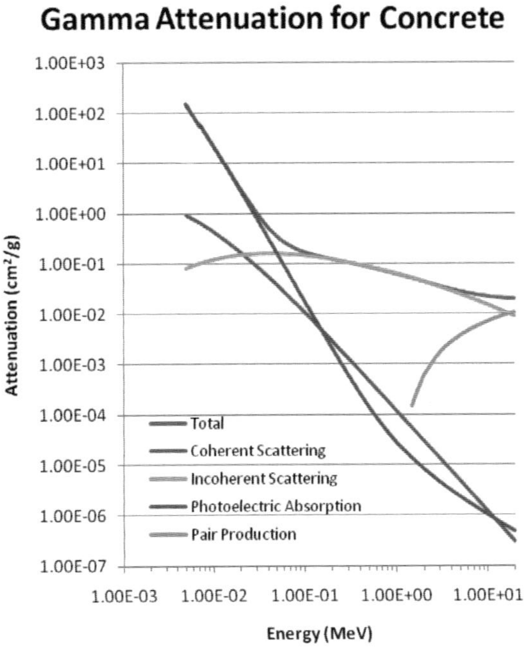

Figure 5. Gamma attenuation coefficients for concrete (Bergen)

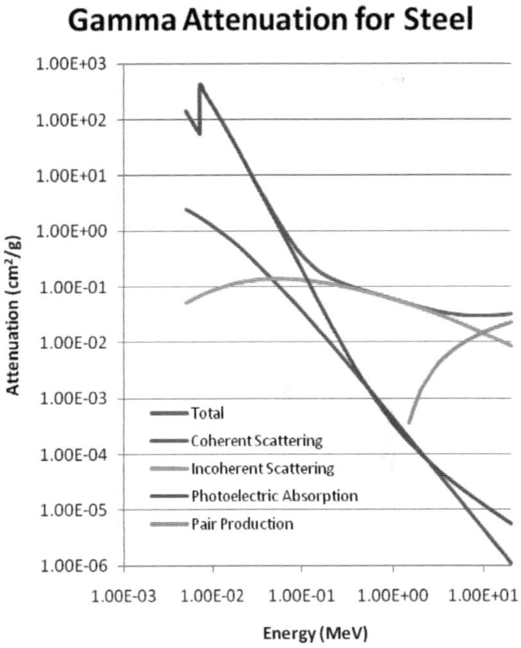

Figure 6. Gamma attenuation coefficients for steel (Bergen)

Neutron Interactions

Neutron cross section data from KAERI for Silicon, Oxygen, Aluminum, and Iron, the primary components of the four materials modeled, are plotted in Figures 7 through 10. These plots identify the total, elastic scattering, absorption, and radiative capture cross sections for neutron energies between approximately 10^{-11} MeV and 20 MeV. In each figure, it is readily apparent that elastic scattering should dominate all other interactions across nearly the entire energy spectrum. Radiative capture plays a much bigger role at lower energies for each

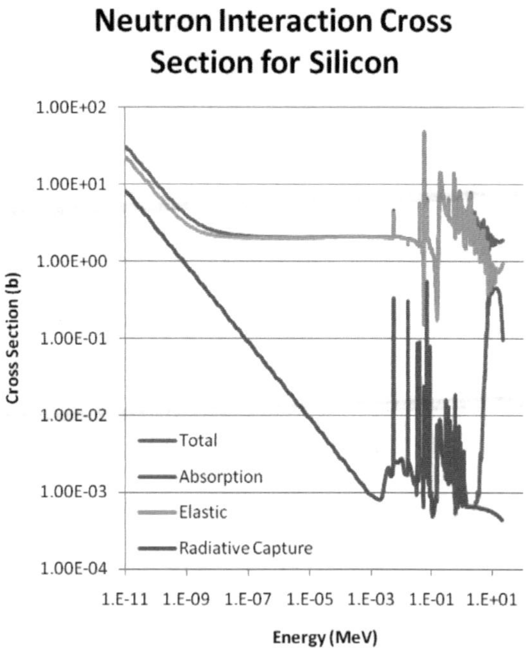

Figure 7. Neutron cross sections for silicon (KAERI)

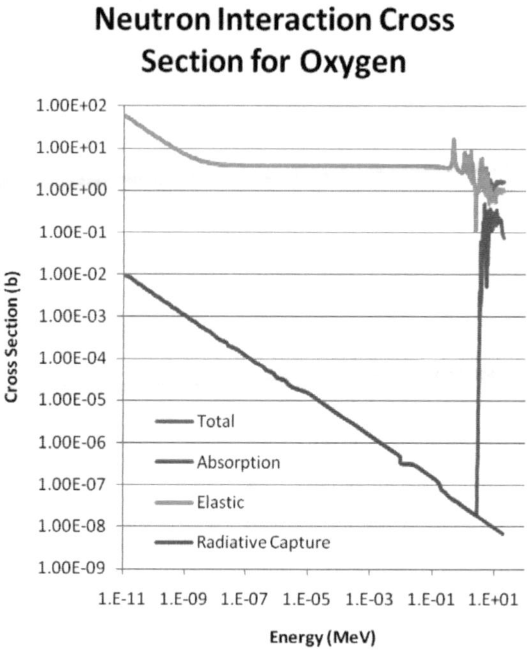

Figure 8. Neutron cross sections for oxygen (KAERI)

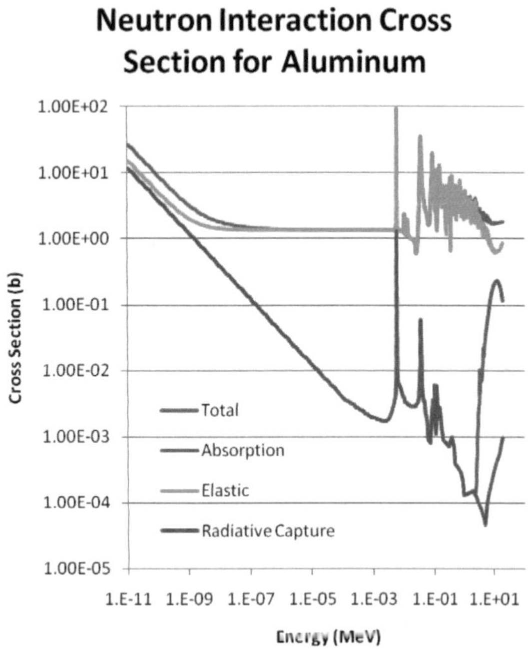

Figure 9. Neutron cross sections for aluminum (KAERI)

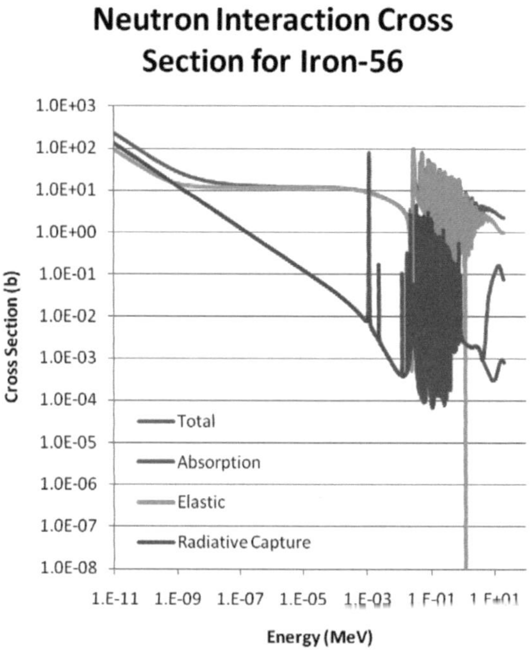

Figure 10. Neutron cross sections for iron-56 (KAERI)

element except oxygen, but resonant peaks in the radiative capture cross section plot between 1 keV and 1 MeV may also rival the elastic scattering cross sections for silicon, aluminum, and iron. Generally, absorption becomes more common at energies greater than 1 MeV.

GEANT4 Model Description

After describing the model inputs, the focus now turns to building the GEANT4 model from these inputs. This model uses a real-world geometric environment as the basis for the simulation, an intense gamma ray and neutron source with user-variable spectral energy distributions as the prompt radiation source, and a detection system which can produce a simulated radiation detector measurement at any point in the world environment. The following section describes each portion of the GEANT4 code which was modified for this forensic analysis case study.

Material Definitions & World and Building Geometries

The DetectorConstruction file in GEANT4 stores the world geometry. This includes the world extents and the "targets" and defines the target material from a list of materials. To construct the three composite materials (A36 steel, stone, and red brick) used in this model, composition data in Table 1 were directly input into the DetectorConstruction file as mass fractions of the composite material.

The world geometry is also defined within the DetectorConstruction file, where the world defines the physical limits of the model space. The "world length" in this model is approximately 600 meters, or about two city blocks in Salt Lake City, producing a 600 meter cube. The world environment is further defined to be composed of air, with its center located at 250 South Main Street, as shown in Figure 4. The world center is assigned the (0, 0, 0) coordinate, which acts as the location reference for all other object geometries.

Lastly, 14 buildings were added as "targets" to the DetectorConstruction file, each with its own shape, dimensions, position, and material. While several object geometries are available within GEANT4, only rectangle objects were used to describe the building shapes. Each rectangle requires dimensions provided as the half-length, which is described in the first three columns of Table 2. The last three columns from Table 2 were used to describe each object's center point, where each target is given a position vector to its center in reference to the world center. These values were all calculated from the geometries stored in the GIS. Each target was randomly assigned a material from the four available composite materials. The results of this process produced the world environment shown in Figures 11 & 12.

Gamma Ray and Neutron Source Description

In order to incorporate a user-defined spectral energy distribution at the radiation source, this model used a GeneralPurposeSource (GPS) to replace the more common ParticleGun. The GPS provides greater flexibility when defining the source characteristics and allows the user to define the characteristics of the source in a macro file accompanying the executable model. In the macro file for this model, the gamma ray and neutron sources were split into two separate isotropic sources with separate spectral energy distributions. Each energy distribution was defined by a histogram using percentages occurring between energies, using the distributions in Figures 2 & 3 as guides.

Figure 11. Finalized world environment with color-coded building attribute descriptions

Figure 12. 3D wireframe representation for the final GEANT4 model

Detection System

The last component to be defined in the model is the detection system. Two options were considered: (1) create a sensitive detector which provides detected gamma ray and neutron spectral energy distributions at any location specified by the user, or (2) generate a scoring mesh to produce large-scale data collection capabilities. The scoring mesh divides the environment into pre-defined cells (in this case, boxes) and counts the particles passing through that cell. The scoring mesh was used in this model at the present time due to time limitations and the ability to produce output visualizations from the environment-wide simulation data.

Defining the scoring mesh takes place in the macro file. A boxMesh is first created to define the scoring mesh. To measure both the neutrons and gamma rays in the world environment, a separate mesh is created for counting neutrons and gamma rays. Mesh commands in the /score directory are then used to define the size of each box and the number of boxes distributed in each direction. Once the mesh is defined, multiple scorers are defined with a filter associated with each scorer. The scorer is a detector, where this detector is defined to count the number of particles passing through it (nOfStep), and the filter specifically limits the scorer to counting specific particles with certain properties. In this case, the filters are used to count either neutrons or gamma rays of different energies, where each energy range is defined in the /score/filter command so that the measurement is divided into a histogram-like distribution. This data can then be output into a text file (CSV) for further analysis.

GEANT4 Model Outputs

Data outputs from GEANT4 can be targeted for multiple uses with multiple programs, but this model specifically uses HepRep and Matlab for output data analysis and visualization. The macro files accompanying the executable file initiate trajectory recording for visualization in HepRep. These visualizations show the path each particle takes from the source and includes information about particle type, energy, and interactions which take place in the world environment for each particle trajectory. Visualizations such as those shown in Figure 13 present some preliminary results from the model.

First, the visualized results show many scattering interactions, matching our expectation after examining the attenuation coefficients and cross section. Ionization taking place at the building surfaces, which is best observed in the left half of Figure 13, likely indicates pair production interactions from higher energy gamma rays. Being able to visualize these interactions is important when considering which areas may require greater attention post-detonation for clean-up and finding good locations for shelters based on the amount and types of radiation present in different areas in the urban environment.

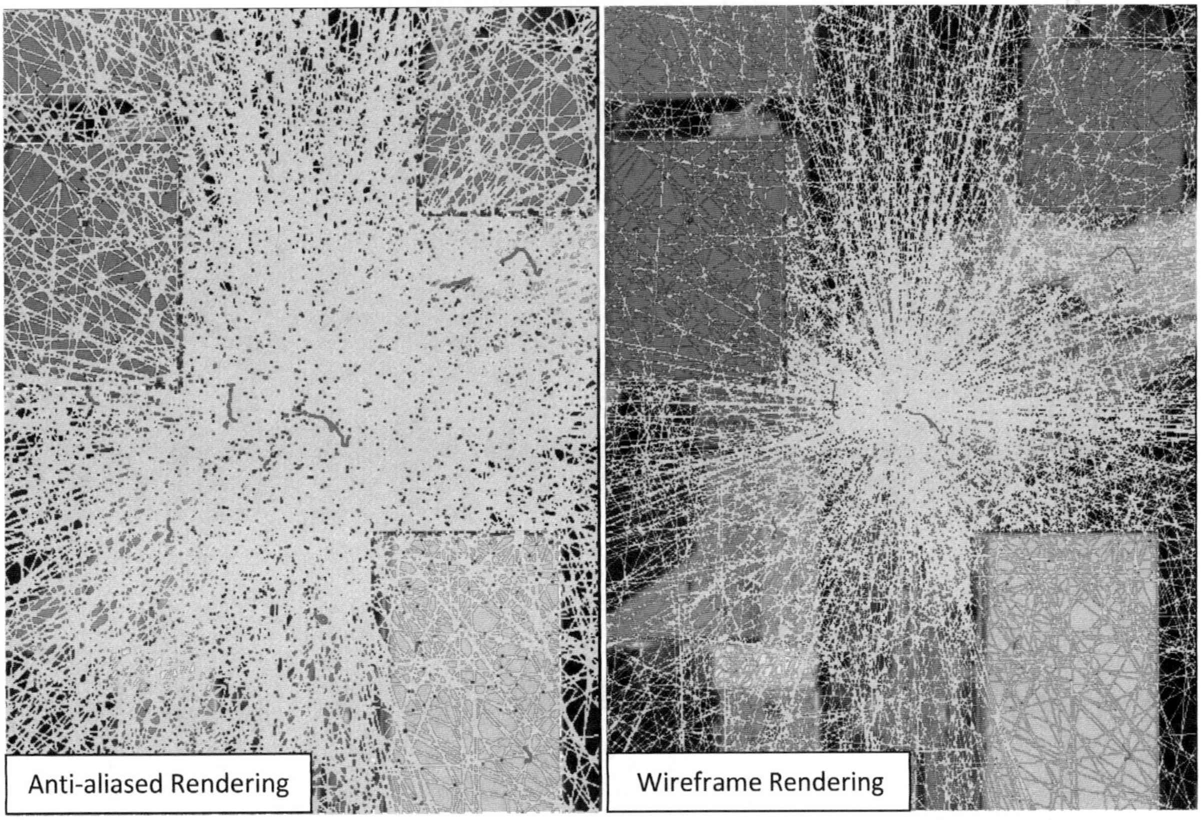

Figure 13. HepRep visualization results overlaid on background aerial photography, demonstrating interactions (left) and attenuation/shielding (right) at building surfaces

From the scoring mesh, Comma-Separated Value (CSV) tables, stored as text files, record the number of neutrons and gamma rays passing through each cell in the mesh. When the data format is converted into matrix form, MATLAB can be used to visualize the number of particles passing through different regions of the entire world environment, as shown in Figure 14. The scoring mesh can be defined with large cell sizes (low resolution results) to reduce run times and minimize data file sizes, but the amount of useful information is quite limited (as shown in Figure 14). However, with higher resolution outputs and more particles running in the simulation, maps such as those shown in Figure 15 could be used to show the effects of building locations on particle movement to a greater degree.

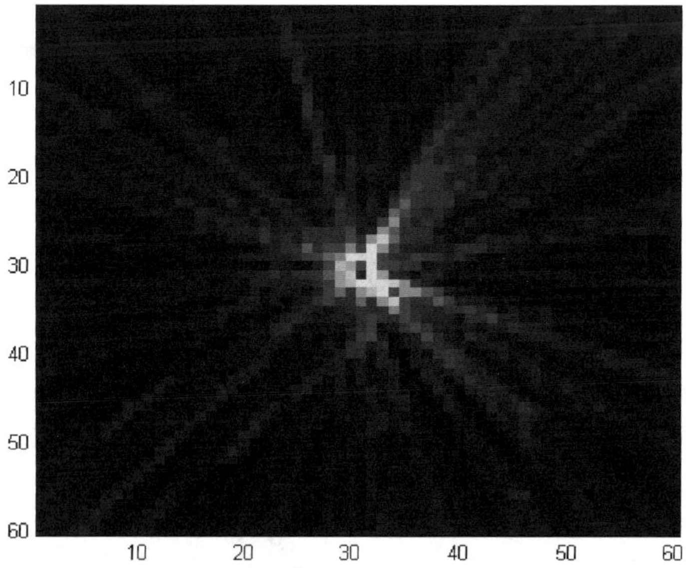

Figure 14. Neutron particle density for the world environment from scoring mesh

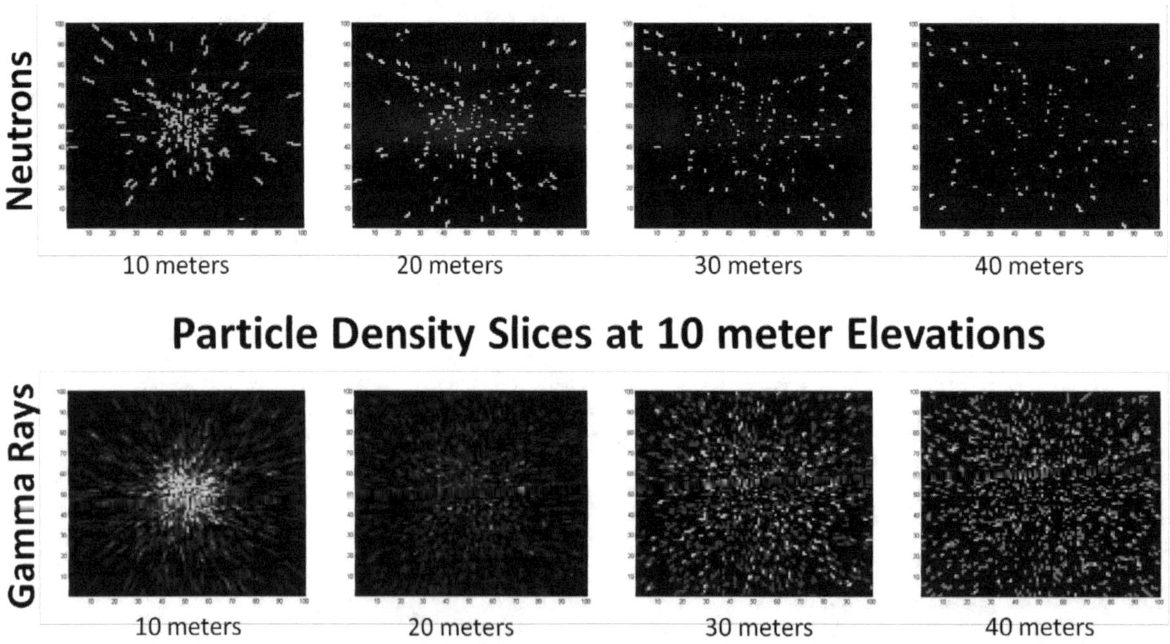

Figure 15. High resolution particle density maps at 10 meter elevation intervals

Figure 15 shows the particle density maps for a very high resolution scoring mesh over the entire simulation environment with a large number of source particles. The neutron density maps appear to show some locations at low elevations with low particle densities, with more random densities observed at higher elevations. This coincides with expected elastic interactions with the building materials, reducing the particle densities in building regions. However, we observe interesting results in the gamma ray density maps. The lowest elevation shows very high densities at the source, and the higher elevations appear to be more random, but the 20 meter elevation map shows an odd reduction in overall density. This is likely due to a coding error or an inconsistent visualization property (i.e. inconsistent color scale) since the effect is not reproduced at higher elevations, where a trend would be observed.

Conclusion

The GEANT4 simulation presented here is a very simplified version of a more capable system that could be implemented in post-detonation nuclear forensics practice. By comparing a measured spectral energy distribution with those estimated by the model with consideration for geographical location and proximity to the source, practitioners may develop better estimates of the source material composition. Also, visualization tools accompanying GEANT4 in HepRep, and the ability to output files which can be used in Matlab for visualization, bring useful tools for examining the effects of different radiation sources in urban environments and the effects of the nearby building environment on exposure just after detonation.

Future Work

Generally, this model is very limited in its current state due to its size, its less accurate building geometry, and its scoring system. As a 600 meter cube, the simulation produces a lot of data because it tracks so many particles in such a large space. By adjusting the limits of the world volume in the detector construction file, the executable could be made to be more efficient by no longer calculating particles below the ground elevation in the area. Similarly, changes could be made in the detector construction file to better define building geometries based upon interior dimensions, or simply using building shells rather than solid objects. Since almost no buildings are completely solid, the current model overestimates the effects the buildings have on shielding and other interactions. Additionally, there is no ground layer in this model, so ground reflection effects are not captured as well. Lastly, the scoring mesh currently employed in the model provides a very simple way of estimating the spectral energy distribution at any point in the world environment, but its low resolution and creation of large data output files hinders analysis. When the resolution is increased to the point where the data is more useful, such large amounts of data are created that it significantly increases run times and produces large CSV files which must still be reformatted for analysis in another application. Using a custom scorer would likely be the best approach because it could be tailored for providing higher-resolution measurements, albeit at specific locations rather than for the entire environment, and better-formatted output files for simpler analysis and visualization.

Acknowledgements

This research was funded by grants from the Department of Energy and Department of Homeland Security as part of the developing nuclear forensics program at the University of Utah.

References

1. American Physical Society and American Association for the Advancement of Science (2008), “Nuclear Forensics: Role, State of the Art, and Program Needs.” Web. <http://iis-db.stanford.edu>
2. Glasstone, S. and Dolan, P. J. (1977), “The Effects of Nuclear Weapons, Third Edition.” U.S. Government Printing Office, Washington, D.C. Web. <http://www.dtic.mil>
3. *Geant4 User's Guide for Application Developers, GEANT4 Materials Database, Appendix Section 9*. European Organization for Nuclear Research (CERN). December 2010, Web. <http://geant4.cern.ch>
4. Blatt, Harvey and Tracy, Robert J. (1997). “Petrology (2nd ed.).” New York: Freeman. p. 66. ISBN 0-7167-2438-3.
5. Punmia, B.C., Jain, Ashok Kumar and Jain, Arun Kr., “Basic Civil Engineering.” Firewall Media, 2003, p. 33. Web. <http://books.google.co.in/books?id=sWZxu_muxylC>
6. *ASTM A36 Steel plate*. MatWeb: Material Property Data. Web. <http://www.matweb.com/search/datasheet.aspx?matguid=afc003f4fb40465fa3df05129f0e88e6&>
7. Berger, M.J., Hubbell, J.H., Seltzer, S.M., Chang, J., Coursey, J.S., Sukumar, R., Zucker, D.S., and Olsen, K. National Institute of Standards and Technology, Gaithersburg, MD. *XCOM: Photon Cross Section Database (version 1.5)*. 2010. Web. <http://physics.nist.gov/xcom>
8. *ENDFPLOT-2.0*. Korea Atomic Energy Research Institute (KAERI) Nuclear Data Evaluation Lab. Web. <http://atom.kaeri.re.kr/ >
9. *Utah GIS Portal*. Utah Automated Geographic Reference Center. Web. <http://gis.utah.gov/>
10. *Buildings of Salt Lake City*. Emporis.com. Web. <http://www.emporis.com/application/?nav=worldmap_ci_bu&id=102671&lng=3>

(213a) Competing Forces On the Extraction of Heavy Metals In Soil Remediation

Cameron A. Grey, Mario Oyanader and Pedro E. Arce, Chemical Engineering, Tennessee Technological University, Cookeville, TN

Heavy metal contaminants found in soil locations pose a tremendous problem to their removal and proper disposal. As to their removal, the extraction of these metals from the ground has been tested theoretically with the application of an electrical field in a soil matrix. This method has been reported to extract specific contaminants with little to no excavation through low permeability soils. In the presence of low permeability soils, these electric fields also theoretical prove to be very promising; however, there is still the need for furthering this technology in modeling as well as in its physical application. A major key in determining the driving forces of this system are the different components of the force (electroosmosis, electromigration, buoyancy, and advection) and how they contribute to the bulk flow of the fluid in the ground. Optimizing this force is one such discovery that will be unveiled as well as the different geometries and the analysis of their benefits. The Debye-Hückel approximation will be employed in order to help solve the differential equations of the electrostatic potential. As a result of this research, these contaminant sites will be readily cleaned with an efficient technology that is easily applied to the area; as well as further the advancement of the use of electric fields in cleaning applications.

See more of this Session: Contaminant Transport and Site Remediation
See more of this Group/Topical: Environmental Division

(450e) Fabrication of Molybdenum-100 Coatings for the Production of Commercial Quantities of Technetium-99m Via the Mo-100(p,2n) Reaction On Low Energy Cyclotrons

Kenneth R. Buckley, Nuclear Medicine, TRIUMF, Vancouver, BC, Canada

Alternative production methods of the most common nuclear medicine tracer, technetium-99m, are currently being explored by many centers in reaction to the uncertainty of molybdenum-99 availability from reactor based fission of uranium-235. While the accelerator production of technetium-99m via the 100Mo(p,2n) reaction has been known for a long time the practicality of producing an adequate supply of the radionuclide with acceptable purity and acceptable economics has yet to be demonstrated. The fabrication of robust thick coatings of molybdenum-100 is a prerequisite to being able to accomplish this. Low energy proton cyclotrons (16-19 MeV) are increasingly available as more health care facilities adopt positron emission tomography and these facilities are potentially a viable source of technetium-99m. However these facilities rarely have the infrastructure to irradiate solid target materials at the high currents required to produce adequate supplies of a radionuclide for local regional use. We are presently implementing the methodology and infrastructure on such low energy proton cyclotrons in order to determine the feasibility to provide urban areas with an independent supply of technetium-99m.

The irradiation energy, irradiation duration, and target purity affect the radionuclidic distribution. Molybdenum-100 is presently commercially available as a metal at 97 to 99.5% isotopic enrichment. The remaining molybdenum isotopes (92 through 99) produce other radio-technetiums via $^{92\text{-}100}$Mo(p,x) reactions that are chemically indistinguishable from Tc-99m. These radionuclides have production cross-sections and half-lives that potentially lead to unacceptable radiation doses to patients. The production of these other radio-technetiums, and the ratio of metastable technetium-99 to ground state technetium-99, are dependent on the irradiation energy and irradiation time. Increasing the irradiation energy will both increase the total amount of technetium-99m produced and increase the radio-contaminants. This effort aims to quantify these contaminants for low energy cyclotron production.

Our efforts are directed toward producing a robust solid target infrastructure that can dissipate up to 5.7 kW of beam power. Targets able to dissipate 4 kW have been constructed by others using a thermal spray method with natural enrichment molybdenum. Targets of sufficiently enriched molybdenum-100 that can be irradiated at high currents and are economically viable have not been demonstrated. We are investigating three general routes of target fabrication onto a chemically inert target backing; electroplating, sintering, and ion beam sputtering of molybdenum. The evaluation criteria for each method include the adhesion to the backing, efficiency of production, purity, and ease of recycling the molybdenum-100 target material.

At the present time thin electroplated molybdenum-100 targets have been fabricated on rhodium backing and irradiated. In order to maximize the purity of the deposit and simplify the recycling of the molybdenum electroplating from an aqueous solution is being pursued. The production of thick layers of molybdenum from aqueous solution is contraindicated by the reducing nature of molybdenum which causes a build up of hydrogen on the deposited surface. Sintering of molybdenum is a known process for the formation of predictable thick layers but the adhesion to a suitable backing remains to be resolved. Initial tests are being conducted by UHV Technologies Inc. (Ft. Worth, TX) to ion beam sputter natural molybdenum. The results of these three methods to produce an economically viable and robust molybdenum-100 target amenable to recycling of the enriched molybdenum will be presented.

Production of ^{99m}Tc on a medical cyclotron. Beaver JE, Hupf HB, J Nucl Med 1971; 12(11):739-741.

Cyclotron production of ^{99m}Tc: Experimental measurement of the ^{100}Mo(p,x)^{99}Mo, ^{99m}Tc, and ^{99g}Tc excitation functions from 8 to 18 MeV. Gagnon K et al., NMB 2011 in press.

Cyclotron production of technetium radionuclides using a natural metallic molybdenum thick target and consequent preparation of [Tc]-BRIDA as a radio-labelled kit sample. Targholizadeh et al., Nukleonika 2010;55(1):113-118.

See more of this Session: Engineering Applications of Radioisotopes for Industry and Medicine
See more of this Group/Topical: Nuclear Engineering Division

(450a) Bulk Production of Generator Parent ^{72}se At the Los Alamos IPF Using a NaBr Target

Beau Ballard[1], Donald E. Wycoff[2], John W. Lenz[3], Silvia S. Jurisson[2], Eva R. Birnbaum[1], Kevin D. John[1], Cathy S. Cutler[4], F. Meiring Nortier[1], Wayne A. Taylor[1] and Michael E. Fassbender[1], (1) Chemistry Division, Los Alamos National Laboratory, Los Alamos, NM, (2)Chemistry, University of Missouri-Columbia, Columbia, MO, (3)John Lenz and Associates, East Lansing, MI, (4)Missouri University research reactor Center, University of Missouri-Columbia, Columbia, MO

Generator radionuclidesconstitute a convenient tool for applications in nuclear medicine. Arsenic-72 (half life26 h) is a medium-lived positron emitting radionuclide for PET imaging that canbe produced as a daughter of ^{72}Se (half life 8.5 d). A ^{72}Se/^{72}Asgenerator system would be suitable for hospital on-site utilization. No portable ^{72}Se/^{72}Asgenerator system is available on the market for convenient, repeated ^{72}Aselution at the point of care ("milking"). Radionuclide generator principles for ^{72}Se/^{72}Ashave been proposed in the literature, including repeated distillation of"grown-in" $^{72}AsCl_3$ from carrier added ^{72}Sestock solutions, electroplating of ^{72}Se as Cu_2Se on Cubackings, and solid phase extraction of ^{72}Se. We describe a ^{72}Se/^{72}Asgenerator system analogous to a well-developed commercial ^{82}Sr/^{82}Rbgenerator system.

Productionof the parent ^{72}Se via 90 MeV protonbombardment of a natural NaBr target has beensuccessfully demonstrated at the Los Alamos National Laboratory's IsotopeProduction Facility (LANL-IPF) following a thorough target safety analysis forhigh beam currents.

Safetyanalysis included a high intensity beam thermal study and product nuclideestimates utilizing Monte Carlo code MCNPX. The thermal study illustrated the highthermal stability with good thermal contact being maintained throughout theirradiation, results will be presented. The ^{72}Se from proton irradiated NaBrtargets was chemically recovered at the LANL Hot Cell Facility in good yield(94%). Good agreement oftheoretical thick yields and experimental results were obtained. Batch recovery of ^{72}As vialiquid-liquid extraction was also effective at 76% yield, with minimalbreakthrough of the ^{72}Se parent (0.9%). Moreover, the feasibility of a solid phasegenerator system will be discussed.

See more of this Session: Engineering Applications of Radioisotopes for Industry and Medicine
See more of this Group/Topical: Nuclear Engineering Division

Nuclear beyond Base-Load Electricity: Variable Electricity and Liquid Fuels

Charles Forsberg
Massachusetts Institute of Technology; Email: cforsber@mit.edu
Proc. American Institute of Chemical Engineers; October 16-21, 2011; Paper 240829

Abstract

The world faces two energy challenges: (1) the national security and economic challenge of dependence on Mideast oil and gas and (2) the need to reduce carbon dioxide emissions from the burning of fossil fuels to avoid climate change. Nuclear energy as a low-carbon domestic source of energy can address both challenges. However, nuclear energy in the United States is only used for base-load electricity production—about a quarter of the total energy demand. To address the two energy challenges, MIT has initiated a series of studies to understand long-term nuclear-renewable energy futures in a low-carbon world. This is a first effort to synthesize what has been learned.

The electricity challenge is to provide variable electricity to match demand. Economics favors nuclear electricity production unless renewable electricity production costs are significantly lower than nuclear electricity production costs. This is because in a low-carbon world the amount of electricity that has to be stored to match electricity production with demand is much smaller in an all-nuclear system than any renewable system. While there are many daily electricity storage technologies to help match electricity production with demand, only two seasonal energy storage technologies were identified: nuclear geothermal heat storage and hydrogen. However, a nuclear renewables electricity system that also produces hydrogen for industrial markets may enable an economic system where a larger fraction of the electricity can be produced by renewable energy sources.

Using nuclear energy to support fossil fuel production can reduce greenhouse emissions relative to current practice but not eliminate greenhouse gas emissions. There is the potential for nuclear biofuels to supply a major fraction of the liquid fuels demand. Liquid fuels from biomass are limited by the availability of biomass. Synergisms between nuclear and biofuels can enable up to three times as much liquid fuel to be produced per ton of biomass. Liquid fuels can also be made from air and water with heat and electricity from nuclear power plants. This option can provide unlimited liquid fuel and places an upper cap on the cost of liquid fuels—2 to 3 times that of the cost of electricity on a unit heat basis.

Key enabling technologies for a low-carbon nuclear-renewable energy system include nuclear-geothermal gigawatt-year energy storage, high-temperature electrolysis for hydrogen production, reservoir heating of heavy oils and shale oil, conversion of lignin (the non-cellulosic component of plants) to liquid fuels, and densification of biomass. Most applications can be met with light water reactors; but some applications require the commercialization of high-temperature reactors.

A nuclear renewables energy future is possible and potentially economic. Nuclear and renewable energy sources have different characteristics and in some systems are synergistic; thus, all-nuclear or all-renewables energy futures are likely to be more expensive and difficult to achieve. Renewables economics are strongly dependent on location—particularly latitude because (1) it drives variable seasonal energy demands and (2) wind and solar inputs are strong functions of latitude. Little work has been done to develop credible low-carbon energy futures for a prosperous world of 10-billion people. The uncertainties are large.

1. Introduction

The world faces two energy challenges. The burning of fossil fuels releases carbon dioxide to the atmosphere with the potential for large changes in climate and pH (acidity) of water and soil. This threatens agricultural productivity because of the coupling between long-term climate and formation of fertile soils implies that changing climate moves agriculture to less productive soils. It also implies rebuilding much of man's infrastructure that is designed for specific climate and sea-level conditions.

The second challenge is dependence on Persian Gulf oil and gas. The four largest oil companies are the national oil companies of Iran, Saudi Arabia, Qatar, and Iraq with respective combined oil and gas reserves of 320, 300, 180, and 140 billion barrels of oil equivalent. For comparison, Exxon Mobil reserves are 15 billion equivalent barrels of oil. Oil prices are political decisions and oil dependence has been a major cause of war. The recent opening of the first global-scale Fischer-Tropsch natural gas to diesel fuel plant by Shell in Qatar (consuming the equivalent of 2.5% of U.S. natural gas production) suggests that within 20 years natural gas prices everywhere will be tightly coupled to oil prices—with the Mideast also determining global natural-gas prices.

As a result of these considerations, we have been examining nuclear-renewable energy futures and what technologies are available—the alternative to the current fossil-fuel-based energy system. This paper examines the technology options and implications with an emphasis on the role of nuclear energy. It is a first attempt to provide a synthesis and the starting point for a path forward.

The energy challenge can be broken into two parts: meeting variable electricity demand and meeting the needs for liquid fuels. The electricity and liquid fuels sectors in the future may be coupled through hydrogen produced using electricity and hydrogen consumed to make liquid fuels.

2. Variable Electricity Production

About 40% of the U.S. energy demand is in the form of electricity. This fraction is expected to increase with time. Some the transport energy demand may be met by electric and plug-in hybrid vehicles. The electricity demand varies with time. Today the variable demand is primarily met with natural gas turbines that have high operating costs but relatively low capital costs. In a low carbon world, this option may not exist. While carbon sequestration technologies may be developed, the cost barrier will be very high for fossil power plants with highly variable electricity output.

Electricity Storage Requirements

One option to meet variable electrical demand is to store electricity when production exceeds demand and use the stored energy to generate electricity when required. We estimated the electricity storage requirements for California under three idealized futures assuming (1) electricity storage systems with no losses and (2) each energy source operated at its full capacity to minimize electricity production costs. In all cases the electricity demand was the real hourly California electricity demand for 8760 hours per year.

- *Nuclear*. All electricity is generated by nuclear plants with steady-state output for 8760 hr/y.

- *Solar*. All electricity is generated by solar thermal trough systems in the California desert using the NREL model of performance and California solar data for 8760 hr/y. The NREL model includes some energy storage in the solar system for daily variations in electricity demand.

- *Wind.* All electricity is generated by wind systems using California wind data for 8760 hr/y and NREL model of wind farm performance. It is unclear if California has sufficient wind to meet its total needs.

The results are shown in Table 1 for all-nuclear, all-wind, and all-solar electricity generating systems. The hourly storage requirement refers to storage requirements based on production and demand analyzing hour by hour for the 8760 hours in a year. The daily storage requirements assume the electricity demand is constant each day (total electricity consumed in one day divided by 24 hours) and the storage system must address variations in electricity demand between different days of the year. The weekly storage requirements assume the electricity demand is constant each week and the storage system must address variations in electricity demand between different weeks of the year. The results of the nuclear storage analysis are similar for most of the U.S. The wind and solar analysis are grid specific because of local variations in (1) wind and solar and (2) electricity demand. In some locations such as California the peak demand is driven by summer air-conditioning loads that partly match solar outputs.

Table 1. Electricity Storage Requirements as Fraction Total Electricity Produced[1, 2]

	Hourly	Daily	Weekly
Nuclear	0.07	0.04	0.04
Wind	0.45	0.36	0.25
Solar	0.50	0.21	0.17

This analysis enables one to understand the different types of storage requirements. For example, one could imagine that the use of a combination of storage technologies (a smart grid, pumped hydro storage, electric cars recharged at night, etc.) could eliminate the hourly variation in electricity demand over a period of one day. However, that would not change the daily or weekly variations in electricity demand and the need for storage systems to handle these longer-term variations in electricity demand. Different energy storage technologies may be required to address hourly variations in electricity demand versus seasonal variations in electricity demand. About half the electricity storage requirements are seasonal.

In an all-nuclear world where the nuclear plants operate with a constant output and provide all electricity that is produced, about 7% of the total electricity must go into storage to provide electricity at times when electricity demand exceeds electricity production. If daily short-term storage technologies are available (smart grid, pumped storage, etc.), this decreases to 4% of total electricity generated. This small fraction is a consequence of the fact that two-thirds of the electrical demand is base-load electricity that matches nuclear power output.

For an all-wind and all-solar California (or elsewhere in North America), electrical storage requirements are much higher. The seasonal electricity storage requirements are 25 and 17% of all electricity production respectively because the sun shines in the summer and the wind blows in the spring. Proposed seasonal energy storage systems have about 50% round-trip efficiency (electricity to storage to electricity)—thus doubling the energy going to storage for real systems. This analysis assumes

[1] I. Oloyede and C. Forsberg, "Implications of Gigawatt-Year Electricity," *2010 International Congress on Advances in Nuclear Power Plants (ICAPP 10),* San Diego, California, June 13-17, 2010, American Nuclear Society (2010) (CD-ROM).
[2] I. Oloyede, C. W. Forsberg, M. J. Driscoll, "Gigawatt-Year Electricity Storage Requirements for Nuclear and Renewable Power Production," *American Nuclear Society Winter Meeting*, Las Vegas, Nevada (November 2010).

a solar trough system that includes storage. There are larger storage requirements for photovoltaic systems. Biomass is storable but the resource base is limited. This leads to several conclusions.

Any low-carbon world requires a replacement for fossil fuels in their roles as energy storage media (coal piles, oil tanks, natural gas storage reservoirs). The scale of energy storage for variable electricity production is massive. For the U.S. it is measured in tens to hundreds of gigawwatt-years of electricity depending upon the source of electricity. This storage capability is orders of magnitude beyond what is possible with pumped hydro storage or other existing technologies. Our analysis identified two possible seasonal storage media: heat and hydrogen as discussed in the following sections.

There are large incentives to develop industrial systems that can use variable energy input on a weekly and seasonal basis—an alternative to energy storage.

Last, in a world with electricity production using only renewable energy sources, electricity production costs have to be a fraction of nuclear power production costs to be competitive because a renewable-only electricity system requires so much more storage capacity. Today wind and solar with few exceptions are uneconomic for the large-scale generation of electricity; but, costs are coming down. The viability of renewables is therefore dependent upon either massive reductions in cost or developing systems with other energy sources and uses of energy so the storage requirements are reduced.

Nuclear Geothermal Heat Storage

A nuclear geothermal system (Fig. 1) uses heat from the reactor at times of low electricity demand to heat a cube of rock about 500 m on a side a kilometer and a half underground and create an artificial geothermal heat source for intermediate and peak electricity production. The heat transfer fluid is pressurized hot or cold water. The technology is based on two commercial technologies: (1) heating of underground heavy oil deposits to lower the viscosity of the oil and thus enable it to be pumped and (2) traditional geothermal electricity production.

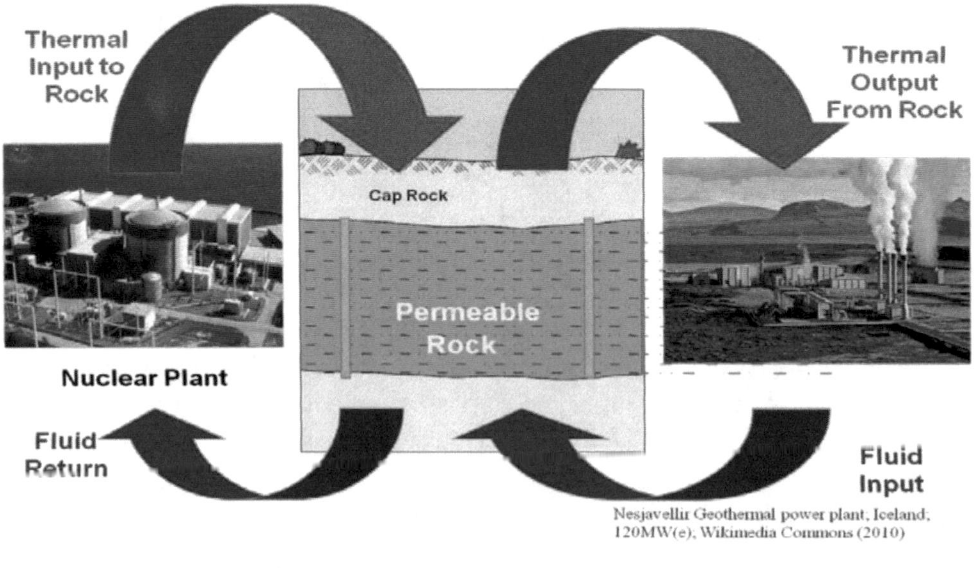

Fig. 1 Conceptual Diagram of Nuclear-Geothermal Energy Storage System

Seasonal heat storage is primarily applicable to large nuclear systems. It is not possible to insulate rock a kilometer underground. However, heat loses vary by the square of the storage system size while heat storage capacity varies by the cube of the storage system size. The fractional stored-heat loses decrease rapidly as the storage capacity increases. For system sizes ~0.1 GW-Y of heat, the losses are only a few percent of the stored heat. The technology only works on a large scale.

An initial analysis was done for the New England electricity grid including an economic analysis (Fig. 2). The assessment indicated that adding nuclear geothermal capacity to the system would reduce the overall electricity generating costs in a system containing (1) base-load nuclear plants producing electricity, (2) base-load nuclear plants producing electricity and heat for the storage system with geothermal intermediate-load electricity production using the stored heat, and (3) natural gas peaking units. The higher capital cost of a nuclear geothermal system versus natural gas turbines makes natural gas turbines preferred for the peak power production—units that operate only a few hundred hours per year. Nuclear geothermal heat storage is a new concept at an early stage of development.

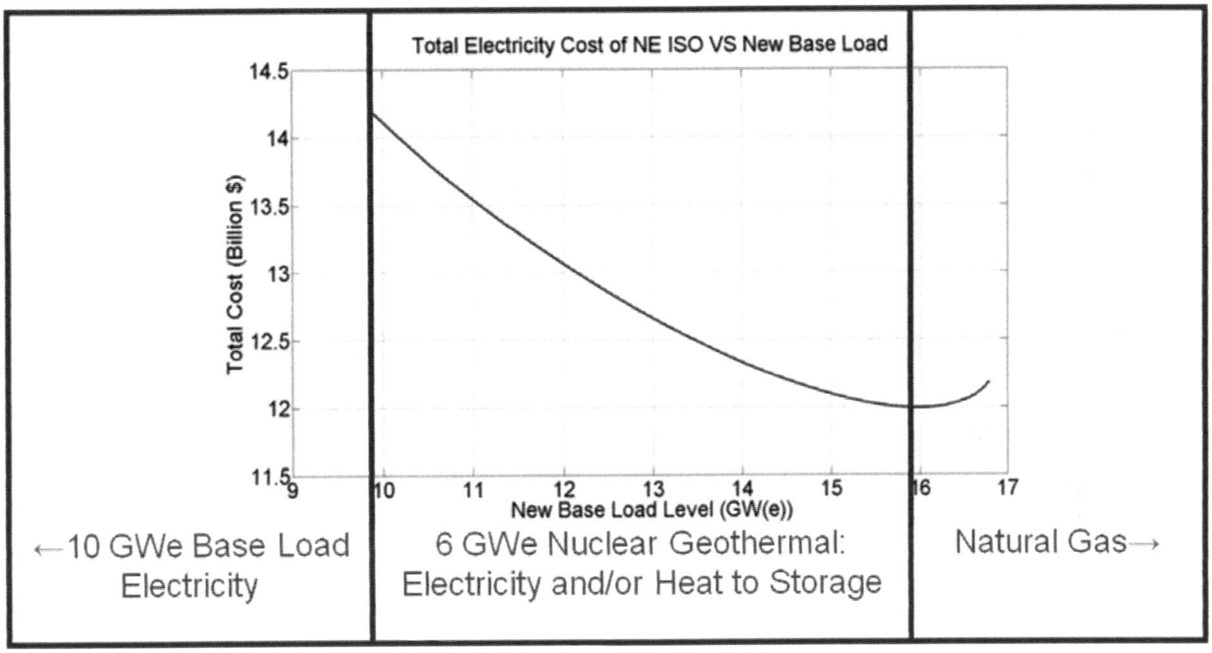

Fig. 2. Total cost of New England Electricity with Increasing Fraction of Nuclear Geothermal Units.

Nuclear Renewables System to Produce Variable Electricity and Hydrogen

Today wind and solar electricity, with few exceptions, are uneconomic without subsidies. However, the production costs are going down. This leads to the question, if the production cost of renewable electricity continues to decrease, is it possible for renewable electricity to become an economic large-scale source of energy? Low electricity production costs do not imply low-cost electricity to the customer unless that electricity is generated when the customer needs electricity. Unfortunately, wind energy production does not match customer demand either during the day or seasonally. Solar partly matches demand for summer air conditioning but there is a massive mismatch with winter demand.

We examined a series of systems and identified one system architecture (Fig. 3) that may enable large scale economic use of renewables for electricity production.[3, 4] A case study was done of a nuclear wind hydrogen natural-gas system to provide the total variable electrical demand for the Midwest electrical grid and hydrogen for the Chicago and Alberta refinery markets. The economics are favorable assuming (1) significant reductions in wind turbine costs, (2) successful commercialization of high-temperature electrolyzers for hydrogen production that can also operate in reverse as fuel cells to produce electricity, and (3) higher-priced natural gas. These are considered reasonable assumptions looking out several decades into the future.

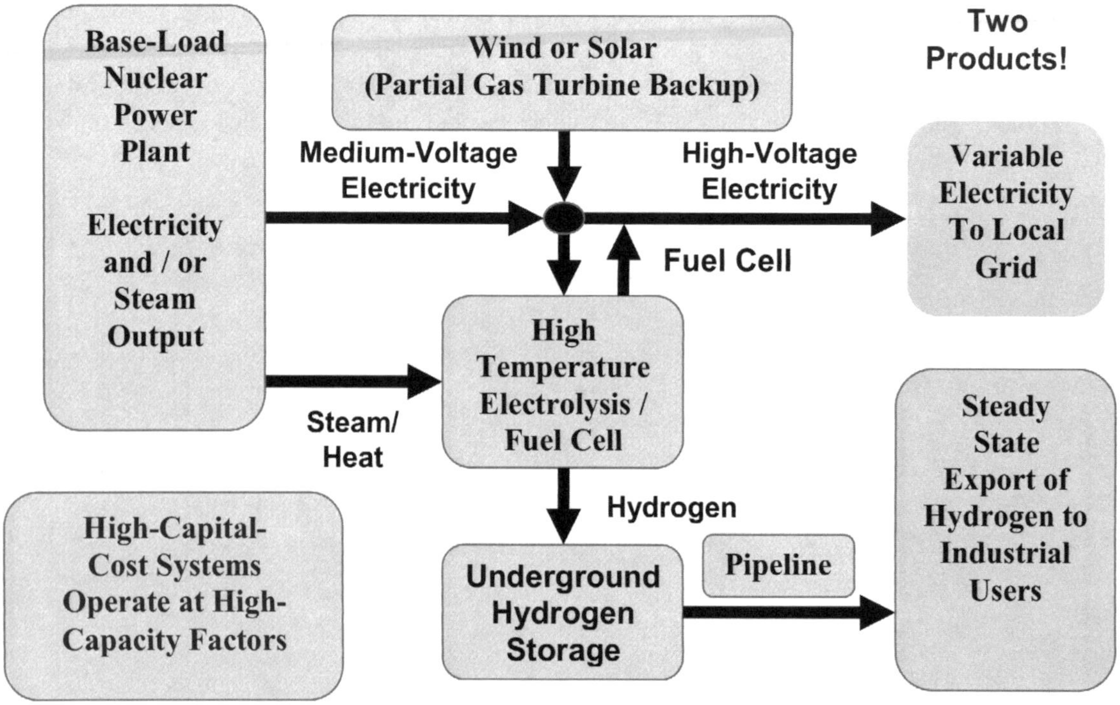

Fig. 3. Electricity-Hydrogen System Design

The goal was an economic system that can meet all electricity demands and produce hydrogen for industrial customers. The primary product is electricity. Economics demands that the capital-intensive low-operating-cost system components (nuclear plants, wind farms, and hydrogen pipelines) operate at near or full capacity while lower-cost components (hydrogen storage, natural gas turbines, and electrolyzers) operate at part load.

The capacities of various components were determined for an optimized system for the Midwest electrical grid that covers parts of the north central United States and Canada using the 2009 hourly

[3] C. Forsberg and G. Haratyk, Nuclear Wind Hydrogen Systems For Variable Electricity and Hydrogen Production. Proc. American Institute of Chemical Engineers Annual Meeting, Minneapolis, Minnesota; October 16-21, 2011

[4] G. Haratyk, C. W. Forsberg, and M. J. Driscoll, *Nuclear-Renewables Energy System for Electricity and Hydrogen Production: A Case Study of a Nuclear Wind Hydrogen System for the Midwest Electrical Grid*, MIT-NES-TR-012, Center for Advanced Nuclear energy Systems, Massachusetts Institute of Technology (June 2011).

electricity demand. This grid has a base-load demand of 39.5 GW(e), an average electricity demand of 61.8 GW(e), and a peak demand of 96.5 GW(e).

- *Nuclear power plants.* Nuclear power plants are only operated under base-load conditions. They are used in two roles.
  - The base-load electricity demand is met with 40-GW(e) of base-load nuclear plants.
  - An additional 2.5 GW(e) of base-load nuclear plants produce a variable mixture of heat and electricity with the heat used for high-temperature electrolysis of water.

- *Wind power plants.* The system contains 50 GW(e) of wind capacity. The annual wind capacity factor assuming 100% availability of the wind turbines is 35.3%. An important characteristic of North American continental wind systems is that there are strong seasonal variations in wind output (Table 2). Peak output is in the spring—the time of lowest electrical demand and maximum hydro output. The lowest output is in late summer—the time of highest electrical demand.

Table 2. Monthly Wind Turbine Capacity Factors in North Dakota (%)

Jan	Feb	Mar	Apr	May	Jun	Jul	Aug	Sep	Oct	Nov	Dec
49	40	42	37	39	22	26	24	35	40	33	37

- *Natural Gas Turbines.* The system contains 25.5 GW(e) of gas turbines to partly meet electricity demands at times of insufficient wind. Backup of wind turbines is also met by fuel cells.

- *High-temperature electrolysis (HTE)—fuel cells (FC).* The system contains 2.5 GW(e) of HTE systems coupled to nuclear plants to produce hydrogen. The HTE system can be operated in reverse as FCs using hydrogen to produce electricity.
  - HTE is the steam electrolysis of water—a process that is more efficient than traditional electrolysis because heat used to make steam partly replaces electricity. The heat is provided from nuclear reactors that produce variable outputs of electricity and steam. A 1-GW(e) nuclear reactor produces 2857 MW(t) of heat. This amount of heat combined with 15,438 MW(e) of electricity (nuclear and wind) will produce 120.81 kg H_2/second. The economic optimization that trades off the cost of added HTE capacity and excess wind capacity results in about 5.5 GW(e)-y of wind electricity being used to produce 1.3 million tons of hydrogen per year.
  - When operated in a FC mode, these units can produce 28.5 of GW(e). The HTE-FC system is optimized for hydrogen production; thus, the assumed efficiency as FCs is only 40%. Hydrogen is more expensive than natural gas because hydrogen is traditionally made from natural gas; thus one would assume all peak electricity demands are met by gas turbines. However, in real electrical grids many gas turbines operate a few hundred hours or less per year to meet peak electricity demands. This is highlighted in Fig. 4 that shows the existing operating hours per year for gas turbines in the Midwest electrical grid. For gas turbines that operate only a few hours per year, most of the cost is capital cost—

not the cost of the natural gas. Consequently, it is more economic to replace 28.5 GW(e) of low-capacity-factor gas turbines by operating the HTE units in reverse as FCs. Hydrogen is a more expensive fuel but the savings in capital costs make this the preferred option. This is a major economic benefit in using HTE that can operate in reverse to produce electricity.

- *Hydrogen pipelines*. Two 36-inch 1000-km hydrogen pipelines move the hydrogen to markets.
- *Hydrogen storage*. Hydrogen is stored underground—an existing commercial technology. It is the same technology used for storing natural gas where in the U.S. the gas storage capacity is about equal to a quarter of a year's production of natural gas.

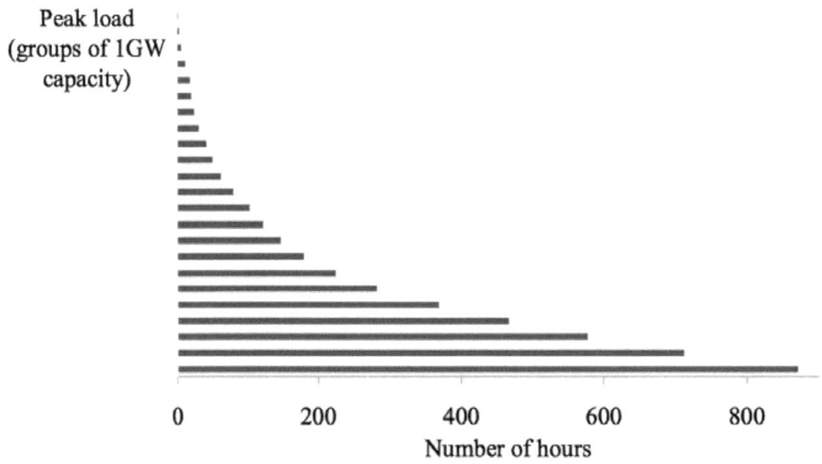

Fig. 4 - Operating hours per year for electricity peaking plants in MISO (2009)

The breakdown of electricity production is shown in Fig. 5. The grid average levelized cost of electricity is \$81/MW(e)-h with a levelized cost of hydrogen of \$3.1/kg and a net hydrogen production of 1.3 million metric tons per year.

Nuclear provides almost 60% of the electricity—roughly matching the two-thirds of the electricity production that is base-load electricity demand. This is driven by the high cost of using wind to match electricity production with demand. The HTE system used as fuel cells produces only 0.5% of the total electricity—again because of the inefficiencies of converting electricity to hydrogen and back to electricity that makes hydrogen an expensive fuel for electricity production. However, the capability to operate HTE as FCs implies large savings in capital costs by replacing gas turbines that operate for only a few hundred hours per year. In this mode, the FCs are a quarter of the grid's non-wind capacity to produce electricity at times of high electricity demand and low wind conditions.

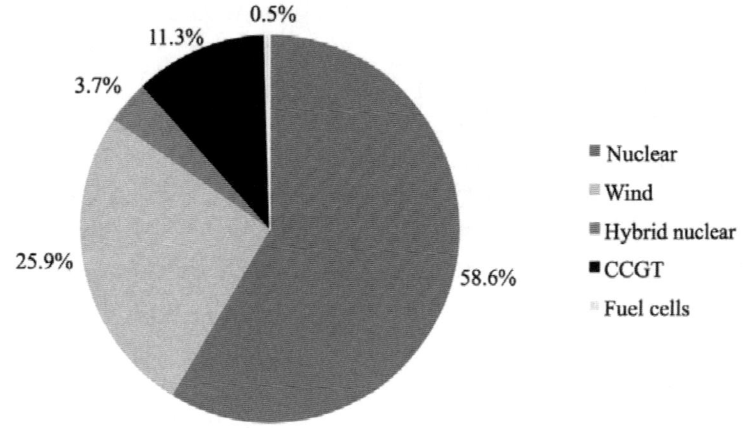

Fig. 5. Electricity generation breakdown

Wind produces 26% of the electricity while combined-cycle natural gas plants produce 11% of the electricity. The system has very low carbon dioxide emissions relative to traditional wind systems. Wind farms typically have capacity factors less than 30% with natural gas backup when the wind does not blow. About 4% of the electricity goes into hydrogen production—primarily for commercial sales. If very low-cost wind systems are successfully developed, the system would optimize toward larger hydrogen production.

Reliable electricity systems using nuclear and wind appear to be viable. The economics depend upon reductions (25 to 50%) in the capital cost of wind farms, commercialization of economic HTE-FC systems ($400/kWe), and higher natural gas prices. The results depend upon assumptions—however our analysis leads us to believe that the system architecture is robust for a wide set of conditions. The low cost of hydrogen storage and the availability of a second market are the enabling factors for the potential large-scale use of renewables assuming major reductions in production costs.

Systems Design

The central challenge of a low-carbon electrical system is the mismatch between electricity demand and electricity production from capital intensive, low-operating-cost nuclear, wind, and solar systems. While there are many types of short-term energy storage systems, we have identified only two seasonal storage media capable of GWY storage. Their characteristics drive the total systems design. Our initial economics analysis indicates that heat storage is likely to be preferred to meet intermediate electrical loads and hydrogen is likely to be preferred to meet peak electrical loads.

3. Liquid Fuels Production

Liquid hydrocarbon fuels (gasoline, diesel, and jet fuel) are the largest energy source today and the primary transport fuel because they have high energy densities that minimize storage volumes, are easy to transport, and are relatively safe to use. There are multiple challenges: economics, national security, and the potential for climate change.

The four largest oil companies are the national oil companies of Iran, Saudi Arabia, Qatar, and Iraq with respective combined oil and gas reserves of 320, 300, 180, and 140 billion barrels of oil equivalent.

For comparison, Exxon Mobil reserves are 15 billion equivalent barrels of oil[5]. Price and availability are political decisions.

Throughout history nations have waged wars to gain control over energy sources. During World War I, the British fleet converted from coal to oil. But the British Empire had very little oil, so the British Indian Army invaded the northern Persian Gulf and in 1920 carved Iraq out of the Ottoman Empire to provide access to oil. On December 7, 1941 the Japanese attacked Pearl Harbor. Japan was short on oil and concluded it had to take over Asian oil fields, but feared the US might block its actions. To prevent that, Japan launched its preemptive attack on Pearl Harbor, precipitating the US entry into World War II.

In 1953 America was drawn into the Middle East when the democratically elected government of Iran responded to a series of oil-company scandals by nationalizing all foreign oil holdings. The United States and Britain reacted by organizing a military coup, overthrowing the Iranian democracy, and establishing a monarchy. In 1979, Iranian anger with the Shah and his foreign backers fueled the Iranian Revolution and the country's anti-American sentiments. US alliances with other Arab monarchs, and with Saddam Hussein of Iraq, have had similarly troubled outcomes. The national security costs of maintaining access to oil are large and increasing with time.[6]

Evidence suggests that the world is reaching the limits of conventional oil production. In spite of high oil prices, production of conventional oil has not increased—something that has not occurred before. This is in line with many predictions made in the last thirty years. It also reflects a world where oil is the primary fossil fuel energy source but is also the rarest form of fossil fuel (Table 3). High oil prices are leading to the largest transfer of wealth in human history and massive trade deficits.

Table 3: World Fossil Fuel Resources[7]

Feedstock for Liquid Fuel	% World Hydrocarbons	Heat Input As Fraction of Heating Value of Liquid Fuel
Oil	2-3%	6-10%
Heavy Oil	5-7%	25-40%
Natural Gas	4-6%	
Gas Hydrates	10-30%	
Oil Shales	30-50%	>30%
Coal/Lignite	20-30%	>100%
Biomass	Annual	To 40%

These factors lead to the strategic question of whether nuclear energy can help eliminate dependence on oil. Liquid fuels production requires a carbon source (crude oil, natural gas, coal, oil sands, shale oil, coal, biomass, or atmospheric carbon dioxide), a source of hydrogen (crude oil, natural gas, coal, oil sands, shale oil, coal, biomass, or water), and energy. Most liquid fuels are made from crude oil because

[5] http://www.petrostrategies.org/Links/worlds_largest_oil_and_gas_companies.htm

[6] http://www.thebulletin.org/web-edition/roundtables/nuclear-energy-different-

[7] C. W. Forsberg, "Nuclear Power: Energy to Produce Liquid Fuels and Chemicals," *Chemical Engineering Progress*, July 2010; M. B. Dusseault, *Cold Heavy Oil Production with Sand in The Canadian Oil Industry*, 2002

its chemical composition (ratio of hydrogen to carbon of two) is closest to that of liquid fuels—making it the easiest and cheapest feedstock to convert into liquid fuels.

An increasing fraction of liquid fuels are being produced from alternative feedstocks—primarily heavy oils and natural gas. These feedstocks are preferred because their hydrogen to carbon ratios are the closest to liquid fuels of the alternative feedstocks; but, it takes much more energy to process them into liquid fuels. Table 3 shows global fossil fuel resources and the heat input to recover and convert various fossil fuels to liquid fuels. In the business as usual future as the world moves to lower-quality feedstocks, greenhouse gas emissions from the production of liquid fuels is expected to rise rapidly per liter of fuel as lower-quality feedstocks are used. This may offset any gains in vehicle efficiencies measured in greenhouse gas releases per person-mile.

There are three options toward a low-carbon liquid-fuels future.

- *Fossil fuels.* Given the importance of liquid fuels, restrictions on greenhouse gas releases are likely to first impact stationary sources of CO_2 releases. That creates incentives to use nuclear energy to provide heat and hydrogen for liquid fuels production so the only CO_2 releases are from burning liquid fuels aboard transport vehicles. There is the longer-term option to remove CO_2 from air and sequester it underground at the same rate liquid fuels are burnt to create a system with zero net additions of greenhouse gases to the atmosphere.

- *Biofuels.* Biomass is a source of liquid fuels with potentially no greenhouse impacts. Plants extract CO_2 from the atmosphere and convert it into biomass that can then be converted to liquid fuels that are burnt. It is a closed loop with no net additions of CO_2 to the atmosphere.

  The primary technical constraint is the availability of biomass. This creates large incentives to efficiently convert biomass to liquid fuels. For the U.S., about 1.3 billion tons of biomass are available for liquid fuels production without impacting food or fiber production. If burnt, the energy content is equivalent to burning about 10 million barrels of diesel fuel per day[8]. If converted into liquid fuels with the biomass providing the feedstock for the biorefineries and the energy for the boilers, the energy value of the liquid fuel is about 5 million barrels of diesel fuel per day. If external sources of heat and hydrogen are available for the biorefinery, the same biomass could be used to produce ~14 million barrels of diesel fuel per day. The importance of biofuels to meet liquid fuels demands depends upon the availability of external energy sources to provide heat and hydrogen to biorefineries so all biomass is converted to liquid fuels—rather than a fraction burnt to provide energy for biorefineries.

- *Liquid fuels from air and water.* Liquid fuels can be made from air and water with sufficient energy input. This provides the ultimate source of unlimited liquid fuels production if energy is available—such as from nuclear reactors. It places a cap on the long-term cost of liquid fuels.

[8] C. W. Forsberg, "Nuclear Energy for a Low-Carbon-Dioxide-Emission Transportation System with Liquid Fuels," *Nuclear Technology*, **164**, 348-367, December 2008.

Fossil Liquid Fuels

Nuclear energy can contribute to fossil fuel production in several ways. The first option is to provide (1) heat and ultimately hydrogen for refinery operations to convert feedstocks to liquid fuels, and (2) heat for oil recovery. Peak refinery temperatures are near 600°C; thus, this nuclear application requires the development of high-temperature reactors. The requirements are well known and there have been many studies on coupling of high-temperature reactors to industrial customers.

The second application is recovery of heavy oil and shale oil—processes that consume 30% of the energy value of the final liquid fuel in obtaining the liquid fuel—not the refining process. Heavy oil recovery involves heating the heavy oil reservoir to higher temperatures to lower the viscosity of the oil until it flows as a liquid and can be pumped out of the ground. Shale oil recovery involves heating oil-shale rock to decompose the kerogen into shale oil and light hydrocarbon gases.

Unlike other industrial processes, heating rock is a slow process requiring months to years because of the low thermal conductivity of the rock. Because of this slow heat transfer, it is not required to heat the rock continuously—one can chose to heat the rock for 16 hours a day and use the heat for other purposes. This characteristic enables the use of nuclear power for simultaneous oil recovery and peak electricity production—a potentially attractive economic option that addresses two challenges simultaneously: variable electricity production and oil recovery.

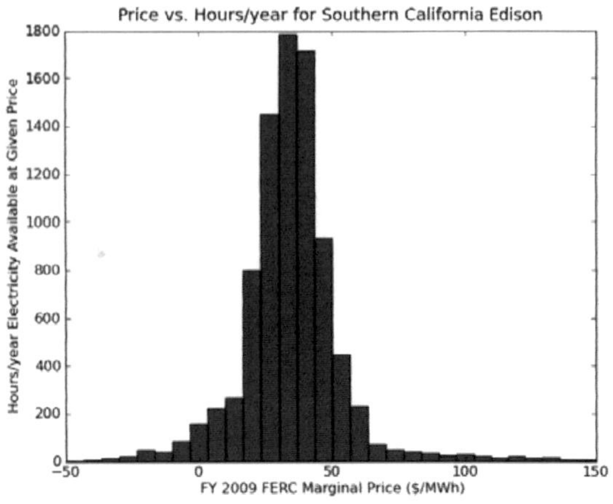

Fig. 6. Southern California Edison Prices for FY 2009[9]

In deregulated electricity markets, the price of electricity varies with demand as shown for the Southern California Edison market in Fig. 6. A nuclear plant could produce heat at night for heavy oil recovery and peak electricity at times of high prices when needed. Our analysis for California heavy oil production suggests a potentially competitive system[10]. This provides a method to maximize the economic return to the reactor owner. It also provides a method to combine nuclear reactors with oil recovery and renewable electricity production to match electricity production with demand. In effect,

[9] CAISO Daily Report Archive. *Federal Energy Regulatory Commission.* [Online] [Cited: June 23, 2010.] http://www.ferc.gov/market-oversight/mkt-electric/california/caiso-archives.asp.

[10] C. W. Forsberg, R. Krentz-Wee, You Ho Lee, and I. O. Oloyede, Nuclear Energy for Simultaneous Low-Carbon Heavy-Oil Recovery and Gigawatt-Year Heat Storage for Peak Electricity Production, MIT-NES-TR-011 (December 2010)

variable heating of oil sands or oil shale is an alternative to energy storage to match electricity production with demand while dramatically reducing greenhouse gas releases from heavy oil production.

Heavy oil production can be done with existing light water reactors while advanced higher-temperature reactors are required for shale oil production. There are some differences in the production technologies.

Heavy Oil Recovery. The principles of heavy oil recovery are simple: heat the reservoir and heavy oil until its viscosity is lowered sufficiently to allow it to flow to production wells. There are three options

The current process is to inject steam underground, heat the rock by condensing the steam, and pump the hot oil out of the ground (Fig 7). The required steam pressure and corresponding saturation steam temperature depends upon the depth of the heavy oil reservoir. At greater depths, higher pressure steam and thus higher temperature steam is required. Some deposits could be heated with steam from LWRs but many deposits would require high-temperature reactors.

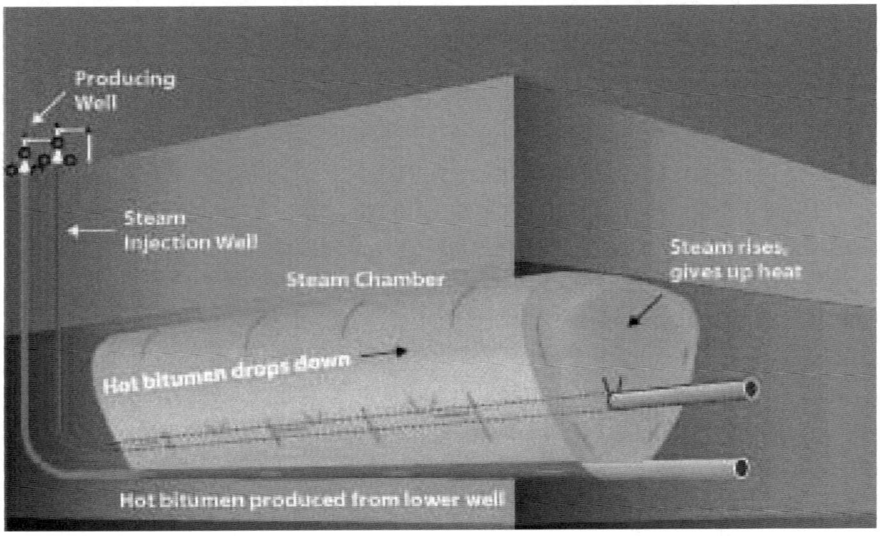

Fig. 7 Steam Assisted Gravity Drainage for Recovery of Heavy Oil from Tar Sands.

The second option is the use of a heavy oil field as a nuclear-geothermal peak power system as described earlier. Because hot water is used to heat the rock, the heat can be provided at any depth by an LWR. One unique feature is that this option may significantly increase the fraction of heavy oil that is recovered and enable heavy oil recovery from deeper heavy-oil reservoirs. Only a small fraction of heavy oil is recovered by traditional heavy oil recovery methods.

- *Heat Injection into Reservoir*. Hot water is injected to heat oil-saturated rock while some oil and cold water is removed at the opposite end of the reservoir. Oil-water separators remove oil from the recovered cold water before this water is heated by an LWR to be injected back to the reservoir to heat the rock.

- *Heat Recovery for Variable Electricity*. Cold water is injected at one end of the operating reservoir while hot pressurized water and oil is recovered. At higher temperatures, the viscosity of the oil is greatly reduced as is the surface tension that holds oil in cracks. Oil-water separators recover the oil from the hot pressurized water before going to the geothermal power plant for peak power production. After going through the geothermal plant, the resulting cold water goes to a second oil-water separator to remove the lighter hydrocarbons that have separated out as the hot water temperature decreased. The cold water is then injected back into the reservoir.

The nuclear-geothermal heat storage facility acts like a washing machine on the heavy oil reservoir over time through many cycles. In such a system, hot and cold water piping would be used to move the location of oil recovery over time. Because of the very large quantities of heavy oil in many deposits, a large reactor could operate its entire lifetime in this operating mode or be converted to only a nuclear geothermal peak power system using the now clean heavy oil reservoir. This application may also be suitable for small LWRs for smaller heavy oil fields.

The third option is using heating pipes (Fig. 8)—a relatively new option made possible by advances in directional drilling technologies. In this case the heat transfer fluid (steam or pressurized water) is isolated from the geology creating a clean heat transfer loop with the option of using an LWR to provide the steam. If heating the reservoir at large depths, an underground pump may be required to return the condensed water to the surface. This option is being primarily developed for shale oil recovery.

Shale Oil Recovery. Shale oil contains no oil, rather it contains kerogen that upon slow heating is converted to a high-quality light oil and various light gases. Recent studies at Idaho National Laboratory[11, 12, 13] have examined using heat from high-temperature reactors in the form of high-temperature steam to heat the kerogen underground. The process is similar to the refinery process of thermal cracking except the carbon residue remains buried underground—a form of carbon sequestration. The heating requirements are about 30% of the heating value of the products thus there are major incentives to use nuclear heat to avoid burning much of the product (usually the light gases) and to reduce greenhouse gas emissions.

The United States has the largest reserves of shale oil in the world. The Green River formation has estimated total reserves of 1.4 trillion barrels, which is greater than the 1.1 trillion barrel of total historical oil production to date. Hence development of this resource would free the U.S. from foreign dependence. However, the oil shale must be heated to ~370°C and thus the need for a high-temperature reactor if nuclear heat is to be used. Fig. 8 shows the leading option to heat oil shale using nuclear heat.

All of these heavy oil and shale oil technologies are in the early stages of development with significant uncertainties. However, together they create potential options to eliminate U.S. dependence on foreign oil with less carbon dioxide releases per liter of liquid fuel produced than from existing crude oils.

[11] E. P. Robertson, M. G. McKellar, and L. O. Nelson, "Integration of High-Temperature Gas Reactors with In Situ Oil Shale Retorting", *15th International Conference on Emerging Nuclear Systems* (15-19 May 2011).
[12] E. P. Robertson, Integration of HTGRs with an In Situ Oil Shale Operation, TEV-1029, Idaho National Laboratory (May 16, 2011).
[13] E. P. Robertson, Economics of Integrating HTGRs to an In Situ Oil Shale Retort, TEV-1276, Idaho National Laboratory (June 22, 2011)

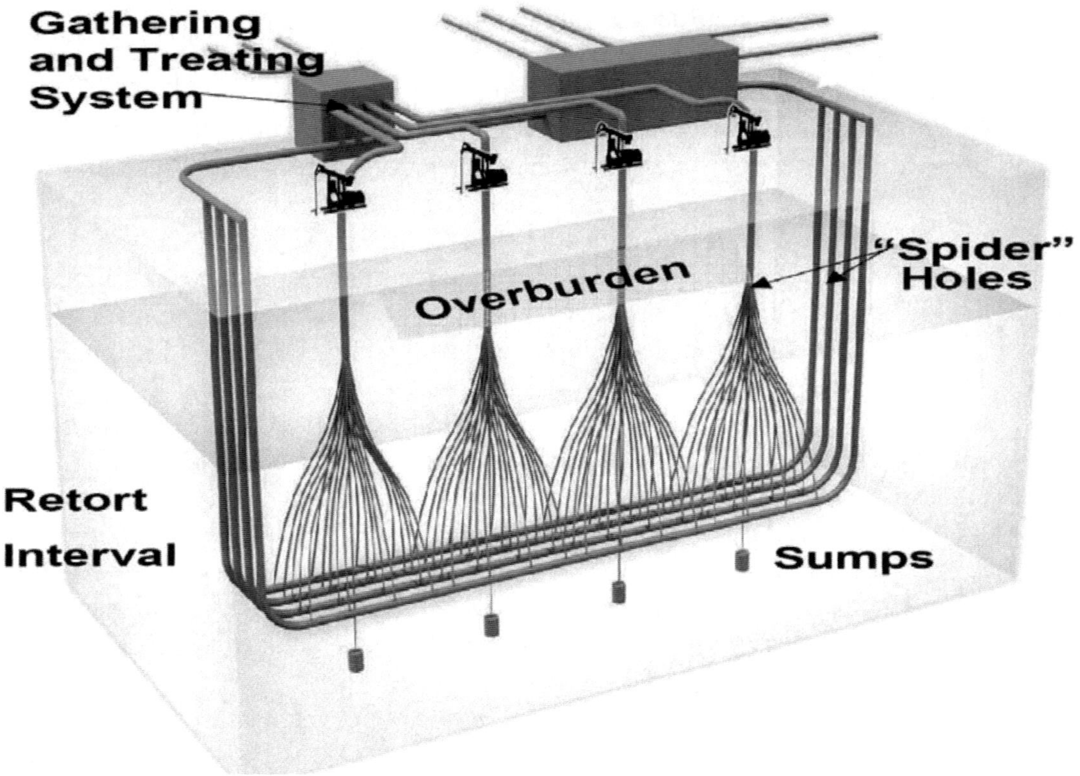

Fig. 8. Isolated Heating System for Heavy Oil or Shale Oil Recovery

Biofuels

Biofuels have the potential to meet most of the U.S. liquid transport fuel needs provided that biomass is strictly a feedstock—not a feedstock and the energy source for biorefineries. Advancing technologies may create within the next decade a large-scale economic biofuels option. Three factors impact the future of biofuels: biomass availability, efficiency in transport of biomass to biorefineries, and conversion processes to liquid fuels.

The first generation biofuels plants converted corn or sugar to ethanol. The next generation biofuels plants will convert cellulosic feedstocks to biofuels because cellulosic biomass (1) is a lower cost feedstock, (2) the largest source of biomass on earth, and (3) is not a human food. The traditional estimates of biomass availability are based on adding up sources of cellulosic feedstocks that are not currently used. That strategy is likely to underestimate total resources.

- *Algae.* Algae growth rates in salt water in deserts can be an order of magnitude faster than traditional crops. If this technology is successfully developed, deserts could be biofuel farms.

- *Restructuring agriculture.* American agriculture, particularly the Midwest Corn Belt, has undergone multiple transformations. It can undergo added transformations to increase simultaneously food and cellulosic production for biofuels—potentially doubling biomass production. The original Midwest agriculture produced biofuels (oats and hay for horses) and various grains. It was restructured to produce corn and meat and latter corn, soybeans, and meat. It can be restructured again for food and biofuels. Options such as double cropping (two crops per year with a summer crop of corn and corn stover and a winter crop for only biofuels production) can dramatically increase both food and cellulosic production—without lowering long-term soil productivity with appropriate recycle of nutrients. There are multiple options[14].

The second challenge is transport. Cellulosic biomass is cheap—the equivalent of $20/barrel oil (current oil prices are $80 to $100/barrel). What is expensive is transporting that biomass to biorefineries and converting it into liquid fuels. The current biorefinery model is direct shipment of biomass to local biorefineries. Because of the low density of biomass, shipping distances may be limited to 50 miles.

There is an alternative strategy. Preprocessing biomass technologies are being developed to have local facilities (equivalent to current grain elevators) pre-process cellulosic feedstocks to produce animal food and densify the cellulosic biomass to the same density as corn. This allows economic long-distance shipment of biomass and if successful has two impacts.

- *Economics of Scale.* There are large economics of scale associated with chemical processing and refineries. Large biorefineries imply lowering processing costs by a factor of two or three.
- *Nuclear Heat.* The heat demands of large biorefineries would couple efficiently with the size of existing large LWRs.

Two classes of technologies exist to convert biomass to liquid fuels: biochemical and thermochemical. Individual facilities may include both technologies to process different biomass fractions. The leading biochemical technologies (Fig. 9) for future cellulosic ethanol plants separate biomass into cellulosic and lignin fractions, convert the cellulosic fraction to ethanol, and burn the lignin fraction to provide the required process heat. The heat input is typically a third or more of the energy value of the liquid fuel. There are multiple competing options[15] that produce different biofuels and in some cases electricity depending upon the feedstock.

In a nuclear biorefinery, heat would be supplied by nuclear reactors so the lignin would not be burnt and could be converted to liquid fuels. Lignin is the biological precursor to oil and coal and thus would be converted into gasoline, diesel, and jet fuel. This implies a much larger output of liquid fuels per ton of biomass. The technologies to convert lignin to liquid fuels are only partly developed.

A distinguishing characteristic of most biofuels processing is the massive demand for low-temperature heat to remove water from biomass and separate ethanol from water. LWRs can provide low-cost heat because of the thermodynamics of electricity production. The value of the steam is determined by its value to produce electricity. Nuclear reactors produce high pressure steam that goes through high, medium, and low pressure turbines to generate electricity. High-pressure high-temperature steam has a high value per kilogram because of its potential to produce electricity but low temperature low-pressure steam (150 psi steam that is used for home heating) has a low value because of its limited

[14] B. E. Dale, B. D. Bals, S. Kim, and P. Eranki, "Biofuels Done Right: Land Efficiency Animal Feeds Enable Large Environmental and Energy Benefits", *Environmental Science and Technology,* **44** (22), 8385-8389 (2010).

[15] M. Laser, H. Jin, K. Jayawardhana, B. E. Dale, and L. R. Lynd, "Projected Mature Technology Scenarios for Conversion of Cellulosic Biomass to Ethanol with Coproduction Thermochemical Fuels, Power, and/or Animal Feed Protein", Biofuels, Bioproducts, and Biorefining, **3**, 231-246 (2009).

value per kilogram to produce electricity. For most of these applications, the nuclear plant would be a cogeneration plant producing electricity from high-pressure steam and sending low-pressure steam to the biorefinery.

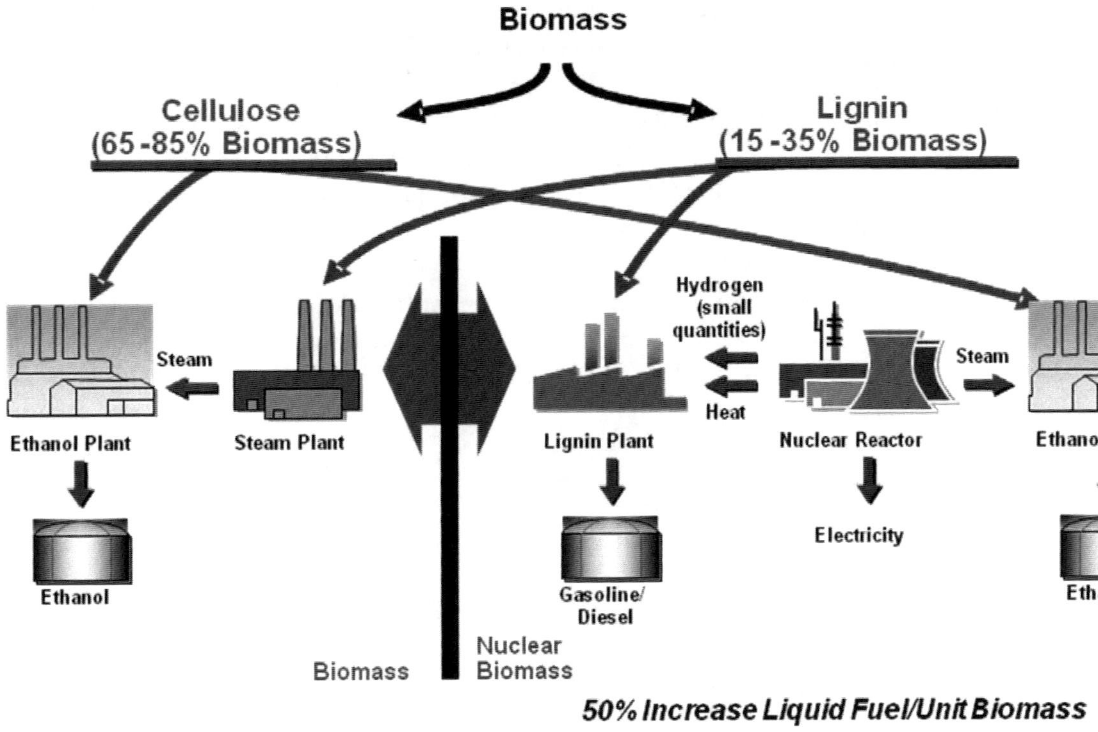

Fig. 9. Alternative Options to Convert Biomass to Liquid Fuels

There are many thermochemical processes from hydrotreating of lignin to conversion of biomass into syngas (a mixture of carbon monoxide and hydrogen) with subsequent conversion into gasoline and diesel fuel using the Fischer Tropsch process—the same process used to convert natural gas and coal into liquid fuels[16]. Some processes require relatively low temperatures whereas other processes require high temperatures and thus high temperature reactors. In all cases, large increases in liquid fuels per ton of biomass would be possible if there was an external source of heat and or hydrogen to power biorefineries.

Liquid Fuels from Air and Water

Liquid fuels can be produced from air and water[17, 18, 19]. We developed flowsheets for production of 6400 barrels per day of JP-5, a jet fuel that is useable as diesel fuel. The feed stocks are carbon dioxide

[16] G. Liu, E. D. Larson, R. H. Willams, T. G. Kreutz, and X. Guo, "Making Fischer-Tropsch Fuels and Electricity from Coal and Biomass: Performance and Cost Analysis", *Energy Fuels*, **25**, 415-437 (2011)
[17] C. Forsberg and J. M. Galle-Bishop, "Jet Fuel from Air, Water, and Uranium" Proc American Institute of Chemical Engineers Annual Meeting. Minneapolis, Minnesota; October 16-21, 2011
[18] J. M. Galle-Bishop, C. W. Forsberg, and M. J. Driscoll, *Nuclear Tanker Producing Liquid Fuels from Air and Water: Applicable Technology for Land-Based Future Production of Commercial Liquid Fuels*, MIT-NES-TR-13, Center for Advanced Nuclear Energy Systems, Massachusetts Institute of Technology (June 2011).

extracted from air and water obtained from the ocean. Heat and electricity to convert the feed stocks into liquid fuel are provided by either a pressurized-water reactor (PWR) or a high-temperature reactor (HTR). PWRs are the most common existing commercial nuclear power technology with peak temperatures of ~285°C. The PWR thermal heat-to-liquid-fuel efficiency is ~22% versus the typical heat-to-electricity efficiency of 33%. HTRs are advanced reactors with peak temperatures of ~700°C. The heat-to-liquid-fuel efficiency is ~31% versus a heat-to-electricity efficiency of 45%.

Because the supply of air and ocean water is effectively unlimited, the flowsheet provides an upper limit for the cost of liquid fuel relative to the cost of electricity—between 2 and 3 times the cost of electricity on a per unit heat basis. On a per unit energy basis, today liquid fuel prices in the U.S. are about equal to the price of electricity.

The study was done in the context of a navy nuclear tanker to manufacture JP-5 at sea at a rate sufficient to supply a carrier strike group under wartime conditions. The flow sheets (Fig. 10) are essentially identical for navy-at-sea or commercial-on-land applications. The major process steps include:

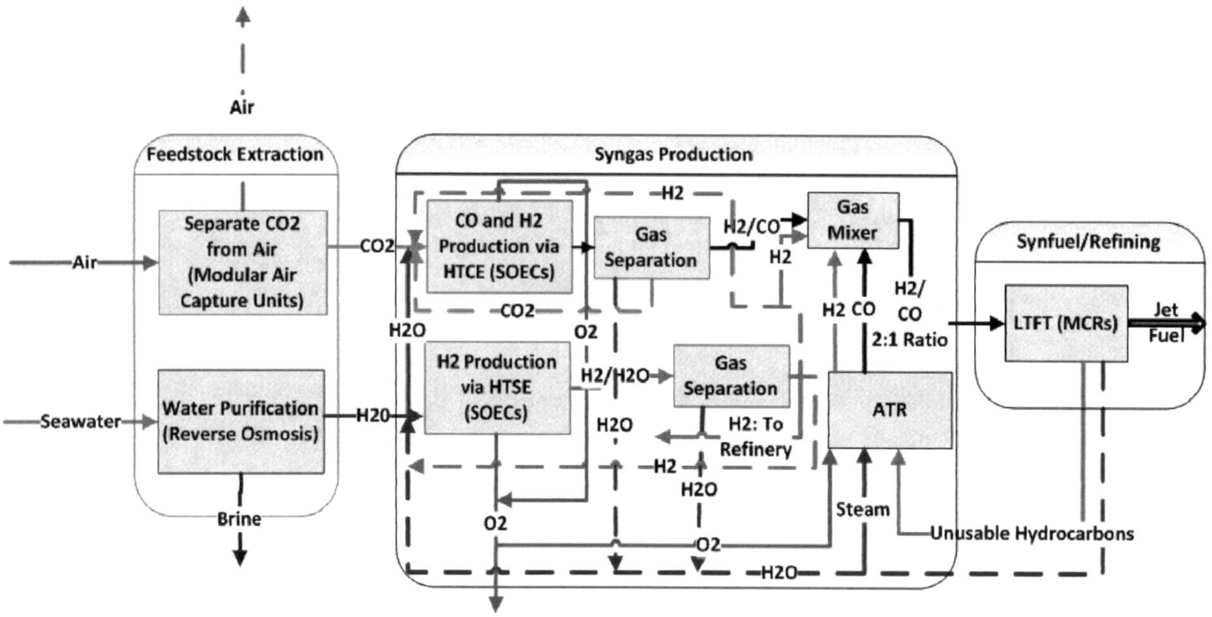

Fig. 10: Top-Level Flow Sheet for Air and Water to Jet Fuel/Diesel

- *Carbon dioxide feedstock*. Carbon dioxide is extracted from air by absorption with regeneration using humid air. This energy-efficient technology is being tested in pilot plants

- *Water feedstock*. Water is purified and converted to steam for use in the high-temperature electrolysis process used to produce hydrogen and the high-temperature co-electrolysis process used to produce syngas.

[19] C. Graves, S. D. Ebbesen, M. Mogensen, and K. S. Lacker, "Sustainability Hydrocarbon Fuels by Recycling CO_2 and H_2O with Renewable or Nuclear Energy", *Renewable and Sustainable Energy Reviews*, **15**, 1-23 (2011)

- *Syngas production*. Water in the form of steam and CO_2 are fed to high-temperature coelectrolysis cells to produce syngas (H_2, CO) and oxygen. The process operates at temperatures near 800°C, is at the pilot plant stage of development, and is based on solid-oxide electrolytic fuel cells operated in reverse. HTE uses about 75% of the energy of traditional alkaline-cell electrolysis. Most of the energy consumption is in this step.

- *Liquid fuels production*. The syngas is fed to a Fischer-Tropsch (FT) system to convert syngas to liquid fuels. Micro-channel chemical reactors are assumed. The FT process using traditional chemical reactors is a commercial process whereas micro-channel chemical reactors are in the pilot plant stage of testing.

- *Refining*. Standard refining technology is used to produce the final JP5 liquid fuel with unusable hydrocarbons recycled through an auto-thermal reformer. The FT process produces a broad slate of products. Producing a mixture of gasoline, diesel, and jet fuel would be easier than just JP5

- *Auto-thermal reforming*. The FT process produces a relatively wide distribution of hydrocarbons. Refining operations can convert many of these products to liquid fuels but there are significant light and heavy products that are not convertible. All of these are recycled back to an auto-thermal converter and burnt with oxygen to produce a syngas for recycle.

There are two potential midterm applications. The first is fuel for military applications such as for aircraft carriers where >90% of the logistical burden is supplying liquid fuels and where the cost of liquid fuels is very high because of the need to protect tankers under war-time conditions. The second is countries such as Iceland with very low-cost electricity and high-priced liquid fuels.

4. Conclusions

Our analysis suggests that it may be possible to create an economic global low-carbon energy system that meets both variable electricity and liquid fuels demands. There remain major uncertainties—but in most cases there are alternative technological pathways. There are several technologies that have large impacts if successfully developed including nuclear-geothermal heat storage, high-temperature electrolysis systems that can operate as fuel cells, lignin conversion to liquid fuels, and densification of biomass. An important transition technology is nuclear-assisted heavy oil and shale oil recovery—a set of technologies that if successfully deployed would eliminate U.S. dependence on foreign oil, reduce greenhouse gas emissions, enable variable electricity production from nuclear power plants by providing a large controllable variable demand for heat (an alternative to energy storage), and enable larger use of renewable energy with variable electricity output from nuclear plants. Economics is critical because the global energy system is a significant fraction of the global GNP—humanity can not literally afford to triple the cost of energy and meet other human needs.

In most applications, light-water reactors can meet the needs. High-temperature reactors are required for some applications. The primary requirements for renewables are reduction in capital costs.

It is relatively easy to envision a nuclear-renewable energy future because of the complementary characteristics of nuclear energy and various renewables—particularly biomsss. The contribution from wind or solar depend strongly upon major reductions in costs. It is very expensive to build an all-renewables world because of the seasonal characteristics of wind and solar imply massive seasonal energy storage challenges. Storage costs can quickly become more important than the cost of the primary energy source (wind, or solar). This is further complicated by the variations in wind and solar over a period of years—similar to the large changes seen in annual hydroelectricity outputs between years due to changes in rainfall (cloud cover, wind, etc.).

Last, what is remarkable is how little analysis has been undertaken to understand what a global low-carbon energy system would look like, what are the primary barriers, and what technologies if developed would have the largest impacts. Energy analysis has historically considered primary energy production technologies (nuclear, wind, solar); however, the viability of such systems may be just as dependent upon supporting technologies that enable seasonal storage of energy and convert energy into liquid fuels.

Acknowledgements

I would like to thank B. E. Dale, M. Driscoll, and R. Wymer for their review of this paper, the many graduate students that did much of the analysis, and Idaho National Laboratory for support of some of this research.

(450f) Production and Application of PET Nuclide 52Fe At Brookhaven Linac Isotope Producer for Radiolabeling of Iron Oxide Nanoparticles to Obtain PET/MRI Probe

Dmitri G. Medvedev, Collider-Accelerator Department, Brookhaven National Laboratory, Upton, NY, Leonard F. Mausner, Collider-Accelarator Department, Brookhaven National Laboratory, Upton, NY and David Schlyer, Medical Department, Brookhaven National Laboratory, Upton, NY

The radioisotope ^{52}Fe is a positron emitter with nuclear properties suitable for imaging applications. Emission of a positron (mean $E_{\beta+}$=340 keV;55.5%) and two gamma-rays 168.7 keV (99.2%), 377.7 keV (1.68%) allows both positron emission tomography (PET) and single photon computer tomography (SPECT) imaging. Production of this isotope can be accomplished in a proton accelerator using ^{55}Mn (p,4n)or ^{nat}Ni(p, xn) nuclear reactions at proton energies above 50 MeV. The use of this isotope has been limited in part due to its short half-life (8.275 h) which makes centralized production and distribution impractical.

Being an isotope of iron, radioisotope ^{52}Fe is ideal for the core labeling of ultrasmall super paramagnetic iron oxide nanoparticles (USPIO). USPIO is a molecular resonance imaging (MRI) contrast agent which appears dark on the MRI image. When labeled with the PET isotope, such as ^{52}Fe, both PET and MRI images can be acquired. In a core labeling approach the radionuclide becomes an integral part of the inorganic center of the NP and, therefore the attachment through the ligands is not needed.

About 40 mCi of Fe-52 decay corrected the end of bombardment (EOB) was produced using ^{58}Ni (p,3p4n)^{52}Fe reaction by proton irradiation of natural nickel foils at 109 MeV at Brookhaven Linac Isotope Producer. Iron-52 was chemically separated from nickel and cobalt isotopes by anion-exchange chromatography. USPIO were prepared by co-precepitation of $Fe^{2+}/Fe^{3+}=1/2$ salts with NH_4OH and stabilized by citrate. Radiolabeling with ^{52}Fe was carried out at pH=6 in sodium acetate buffer for 30 min at room temperature. The labeling yield of 78% was achieved. The obtained nanoparticles were used to obtain MRI and PET images.

See more of this Session: Engineering Applications of Radioisotopes for Industry and Medicine
See more of this Group/Topical: Nuclear Engineering Division

(498a) State-of-the-Art Techniques and Challenges in Characterization of New Hydrogen Storage Materials

Andre Levchenko, Setaram Inc, Pleasanton, CA

The new hydrogen storage materials discovered in the last few years present a number of challenges to the research community from a viewpoint of materials characterization. These include both evaluation of the performance in real-world applications of hydrogen storage materials and understanding the underlying fundamental mechanisms controlling their properties. This presentation gives an overview of state-of-the-art characterization techniques for hydrogen storage materials. It covers simultaneous TPD-gas sorption/mass-spectrometry measurements, volumetric PCT isotherm measurements on thin-film and extremely small metal-hydride samples, as well as direct van't Hoff and in-situ calorimetric measurements of enthalpy of formation of hydrides. Challenges in performing measurements on particular types of materials or under demanding conditions are discussed.

See more of this Session: Aspects of Hydrogen Storage Materials
See more of this Group/Topical: International Congress on Energy 2011

(343d) Chemical Engineering in Nuclear Separations

Terry A. Todd, Idaho National Laboratory, Idaho Falls, ID

The processing of used nuclear fuel and radioactive waste for improved waste management options requires the involvement of a number of technical disciplines. Research, development, scale-up and full-scale implementation of unit operations require significant involvement of chemical engineering expertise. This talk will discuss the importance of the chemical engineering discipline in support of these activities and the need for increased involvement of, and support for, chemical engineering faculty and students.

See more of this Session: Wilson Award Plenary Session on Nuclear Chemical Engineering (Invited Papers)
See more of this Group/Topical: Nuclear Engineering Division

AUTHOR INDEX

American Institute of Chemical Engineers
3 Park Avenue
New York, New York 10016-5991

CURRAN ASSOCIATES INC
proceedings.com